13

During the 1960s, vessels like this fractionating column were virtually churned out by Head Wrightsons. They were still a spectacular sight as they left the Trafalgar Street works when being delivered. KVN 860E is seen heading up this load bound for ICI North Tees, Seal Sands, on September 19 1968 using the Crane solid tyred bogies.

MOVING MOUNTAINS

An illustrated history of heavy haulage

Bob Tuck

PSL Patrick Stephens, Cambridge

Acknowledgements
The author should like to thank all those who have given permission for their photographs to be reproduced in this book. They are, in alphabetical order: *Bristol Evening Post*, British Petroleum, Central Electricity Generating Board, *Daily Mail*, General Electric, Head Wrightson Teesdale Ltd, *Hendon Times*, Tom Llewellyn, N.E.I. Parsons, *Newcastle Chronicle and Journal, Stafford Newsletter*, Alan Timbrell (British Airports Authority, Gatwick), John Wynn, *Western Mail and Echo*.

Others who have helped with research are: Roy Brandley, John Brown, Ted Fitzpatrick, Tom Llewellyn, Alan Simpson and John Wynn. Last but not least are Rick and Dennis without whose help this book would still be in its infancy.

First published in 1983

British Library Cataloguing in Publication Data

Tuck, Bob
Moving mountains.—(World trucks)
1. Commercial vehicles—History
I. Title II. Series
629.2'24'0904 TL230

ISBN 0-85059-614-9

Photoset in 10 on 10 pt Baskerville by Manuset Limited, Baldock, Herts. Printed in Great Britain on 100 gsm Fineblade coated cartridge, and bound, by R.J. Acford Limited, Chichester, Sussex, for the publishers, Patrick Stephens Limited, Bar Hill, Cambridge, CB3 8EL, England.

Also available from PSL—World Trucks Series
No 1: ERF
No 2: Scania
No 3: Seddon Atkinson
No 4: MAN
No 5: DAF
No 6: Dennis
No 7: Volvo
No 8: Scammell
No 9: Fiat
No 10: AEC
No 11: International
No 12: Berliet
No 13: Magirus
No 14: Leyland

Titles in preparation
No 15: White
No 16: Thornycroft & Guy
No 17: Ford Europe
No 18: Mercedes Benz
No 19: Peterbilt
No 20: Albion & Morris
No 21: Foden

Contents

Scammell Contractor—fear of nought.

Left Horse power of the original kind, apparently double headed, seen about to leave John Thompson's at Wolverhampton in the 1890s. The markings on the trailer tend to suggest the outfit was owned by the local railways.

Below Although this photograph was taken in Australia, and as late as 1921, it does show the size and difficulties of the large horse teams which were needed before the introduction of the powerful steam traction engine.

1. In the beginning there was Norman E. Box

The movement of heavy objects from one place to another has been a challenge to man since the beginning of time. The Pyramids and Stonehenge are both famous results of some very heavy hauls, but when it became socially unacceptable to use human slaves, man had to harness the power of the animal to cope with his heaviest load. Even the carriage of goods by barge on the sleepy canals depended on the pulling horse but, with the operation of George Stephenson's most famous invention, the vogue of hundreds of years of transportation was soon to change. 1825 saw *Locomotion* preceded by a man waving a red banner, but the success of the *Rocket* and the Liverpool to Manchester railway in 1829 soon saw the disappearance of the escort, although he did not remain unemployed for long.

In 1801 Richard Trevithick had invented a steam road carriage which could outpace the traditional horse-drawn ones and in 1803 one of his steamers ran in the streets of London. Fellow Cornishman, Sir Goldsworth Gurney, developed this concept and in July 1829 his carriage went from London to Bath and back at a speed of 15 mph. The steam carriage was set to take off but for the political clout which the established stage coach operators and railway owners could muster. Leaders in the commercial field were a well known firm of carters called Pickfords, who had a well organised system similar to a modern day freightliner in operation with the railways carrying their carts nationwide whilst the horses finished the haul from the terminals. To allow the steam wagon in would destroy this business. Legislation created heavy tolls for steam-driven road vehicles and, in 1865, the law enacted a general speed limit for mechanical road vehicles of 4 mph which was further reduced to 2 mph in towns. If this was not oppressive enough, every such vehicle had to be controlled by three persons, one of whom had to walk in front carrying a red flag.

Strangely, the railways failed to take advantage of this virtual monopoly preferring to squabble amongst themselves rather than concentrate on promoting an ideal conception of transport. It took 20 years before a standard gauge was agreed and the 19th century was one continuous conflict between rival companies. Territorial boundaries were closely set and fiercely guarded but had effort been put into making platforms lower, tunnels wider and higher, plus providing more space between opposing lines, then the question asked a century later, 'Why doesn't it go on the railway?' would never have been considered.

Off the rails, horse teams increased in number to haul the ever-growing size of load created by industry, although the four-legged beast was not without problems. Harnessing the power of 20 or 30 horses to pull in concert was difficult enough, but one thing which could not be anticipated was their natural discharge of waste. So much of a problem was it that the manager of one large north-eastern heavy engineering business said as he surveyed his strewn site, 'We've got that much horse dirt in the factory, I don't know whether it will be easier to move the dirt or to move the factory'.

Unfortunately for rose growers, 1896 saw the easing of legislation and combined with the development of the internal combustion engine, the truck as we know it was born. Heavy haulage was still the province of the steamer, the power of the traction engine being unsurpassable in this field. Numerous haulage contractors had the odd one or two big locomotives to pull the occasional heavy load, but the father of heavy hauling must have been a man called Box. Not for him the rapid trunk service into the metropolis, plenty could do that. Eight or ten tons was not worth wasting his time over, but if you wanted to move 30, 50 or 70 tons—then send for Norman E. Box, the heavy haulage specialist. From his base at Ardwick, Manchester, his big Fowlers clanked their way all over the country. Ships' boilers, electrical transformers, castings, anything with weight, anything which would move, then Norman E. Box would move it. If he could not do it, it just could not be done.

They never travelled very fast but the Box road trains must have been an impressive sight. Hitched up to the big locomotive hissing steam and belching smoke, was the load-carrying trailer. A very basic piece of equipment, solidly made, a low slung platform having four, six or even eight metal wheels. No rubber yet in these early days, just solid steel straight on to the roadway. Next in line was the tackle wagon, pulling in turn the coal-carrying trailer which hauled the living van. Yes, four trailers gaily trundled along in the path of their leader and although Lancashire may have been the base, it was Scotland, the North-East, the Midlands and the South where the Fowlers worked. The drivers, as can be expected, were hard men. Black with coal dust, scorched with fire and steam, they worked hard but they also played hard. On a good day with a heavy load, 15 miles was all that you could manage to do but, as coincidence would have it, the journey always seemed to end at a welcoming hostlery. Copious quantities of ale were consumed in another cloud of steam and often the drivers had to sleep the effects off in the steamer's coal bunker as they were too drunk to find the living van. Next morning, bright and early, they prepared to make a start. It was the driver and his hangover who saw to the fire, the mate to the breakfast. After removing the muffler plate, on top of the chimney, which had ensured that the fire did not burn itself out, the raking, stoking and building up meant that in half an hour the boiler was blowing off steam and itching to be at work. Having dined invariably on steak and eggs, progress was recommenced but if that regulator was opened too harshly, the front wheels of the loco would leap four or five feet off the ground as the Fowler gave its best. Names like *Atlas, Talisman, Titan* and *Rover* were proudly affixed to reflect the strength of the vehicle but there was to be one name with an entirely different connotation.

All haulage contractors, from time to time, came into contact with the police who were endeavouring to uphold the law of the land. But getting done for doing 9 mph when your speed limit was 5 mph, or 14 mph when your limit was 12 mph, just seemed too much. The tail light, being a storm lantern, did on occasion go out but why did the police have to issue a summons every time? Box finally rebelled and decided that, no matter what it cost, he would get the finest lawyer in the land and fight. Mr Joynson-Hicks was the best and lived up to his reputation by being able to create just enough doubt about each and every case to bring a 'not guilty' verdict. Justice at last. The law may not have been too pleased but Box was and to show his appreciation the name *Jix* was affixed to one of the Fowlers—power of another kind.

Having substantial contracts for moving transformers in the fast developing electrical industry meant that the Box steamers were kept busy. The railways could haul them so far, but from the railhead to site was Box's domain. It was not unknown when a crew rang in from, say, Glasgow to receive details of their next job for them to be told to run down to Bournemouth. 450 miles empty was a hard slog for ten days at a maximum speed, when coasting down the likes of Beattock and Shap at a hairy 20 mph. It took four revolutions of that little wheel before the slack in the steering chain was taken up, so ensuring the directional stability of the outfit even at this leisurely pace took a lot of nerve and skill. Running loaded was entirely different. Every decline saw the mate leap out to attend to the wind-on trailer brake, the only assured manner of slowing the outfit down. By shouts and gesticulations the driver could tell the athletic mate running alongside the trailer to apply more or less brake. It was a change for him from toting sacks of coal to keep the bunker full and then every seven miles having to divine for water. Not for the steamers strategically placed water tanks on the side of the road, but more a situation where any pond, stream or river assumed far more importance than the modern day filling station.

Box was not alone in the heavy haulage field and in the early 1920s the competitors were given the opportunity of closing the gap in the standard of service which could be offered. True, when compared to the Fowlers, they were tiddlers but the new Scammell articulated machinery transporters were an exciting step forward in heavy haulage. An enclosed cab and a windscreen, positive steering and rubber tyres, albeit solids, were real luxuries. Norman Box bought four or five of these for the small 12-15-ton loads and they were also snapped up by people like Wynns, Marstons of Liverpool, Rudds and the mighty Pickford of London. This old established carter had taken a long time to give up its horses but on entering mechanisation it had gone into it with great enthusiasm. Being a large company it was susceptible to the whims of big business and in 1920 was acquired by the Hay's Wharf and Cartage Company as a wholly-owned subsidiary. The name and identity remained, as did an increasing awareness in the importance of heavy haulage, a big change from their previous long-standing 'smalls' business.

Marston Road Services was fighting hard to overcome the virtual Box monopoly in ultra-heavy haulage which had created a fair degree of financial hardship to any competitors. To harden this fight they gambled hard by putting to work in 1929 one of the most famous vehicles in the business. KD 9168 was the first articulated vehicle able to carry a genuine 100 tons, a big brainwave from the Percy Hugh and O.D. North combination of Scammell. The prime mover had a large riveted frame and solid tyres all round, but what was particularly interesting was that the rear axle had four pairs of these rubber tyres mounted in a line. The transmission from the Scammell four-cylinder, 7-litre petrol engine, which developed a massive 86 bhp, led via the remotely mounted eight-speed gearbox to a countershaft mounted midway along the chassis. Heavy-duty chains then took the drive from the end reduction boxes to the large rear wheels. Three fuel tanks were necessary because of the 1 mpg consumption rate at a flat out 5-6 mph, the mate being kept busy having to hand pump the petrol up to the centrally mounted scuttle tank. The semi-trailer was a hefty double girder construction braced to the gooseneck by a pair of hydraulic struts with screw-jack locks. This gave the crew the ability to lift the load as much as 15 ins above road obstructions or to extricate the tractor's driving wheels from the depths to which they sank in the highway. The latter was a problem which was to remain with this beast for the rest of its life. Perched at the extreme end of the trailer was a small hut, the home for the rear bogie steersman. It was his job to control the direction of the 16 solid tyres supporting the back end and, to aid communication, a telephone link was fitted giving direct contact with the driver. In this one outfit an advance was made in heavy haulage motor vehicle design the like of which has never been seen again.

Norman E. Box could withstand the publicity of this acquisition by MRS but what he could not stand, in this time of recession, was the pressure inflicted by the Hay's Wharf Cartage Company and it, too, sold out and became a subsidiary, but once again retaining its name and identity. To Pickfords this meant that instead of being competitors of 'big Box' they were now partners, so with an interchange of staff at all levels some of the secrets of heavy transportation were quickly shared. Box remained committed to the old-fashioned steamers for the heavy work although the trailers had become a great deal more sophisticated. The load carrier manufacturers in this era included Dyson of Liverpool and Eagles of Warwick. Even Fowler themselves produced a very strong double-cranked low-loading trailer on rubber solids for Box, but the strongest trailer of the day came from Cranes of Dereham in Norfolk. Although strictly speaking it was two separate bogies, the 64-tyred 32-wheeler was normally run as one massive trailer with a rated capacity for 110-ton loads. The axles were arranged to oscillate about a longitudinal axis so that the wheels followed any uneven road undulations. 19 in by 3 in wind-on wheel brakes were fitted and when running together a pair of telescopic drawbars ensured that the rear bogie followed in the track of the front one. Cranes had obviously done their homework with this trailer for, 40 years later, their similar solid bogies were still the recognised way of

An interesting configuration of trailer is seen about to leave Trafford Park in 1906 supporting this large iron casting destined to be fitted in the ship *Lusitania*.

carrying ultra-heavy long vessels demanding low carriage because of their high centre of gravity.

The Box company were obviously impressed with the performance of their little Scammells and of MRS's 100-ton artic so, when BLH 21 came up for sale on the second-hand market, they snapped it up. It had started life as nearly identical to KD 9168, but on completing its first journey to Cornwall, the semi-trailer was left behind as a base for the stone-crushing machine it was carrying. Joining a fleet of heavy locomotives hauling drawbar trailers it seemed natural to stay as a ballasted tractor and proved more than equal to the Fowlers. In this form very little weight was transmitted on to the front axle so it, too, copied the starting leap of the steamers thus, when one of the drivers nicknamed the vehicle *Leaping Lena*, it seemed so apt that it stuck for life.

Heavy haulage in this time was real pioneering. It was not unknown even for the experts to encounter a low bridge on their route which just would not let the load through. No scratching their heads for these lads, there was no alternative to unloading on to the roadway, towing it under the bridge on skids, then re-loading the vehicle on the other side. Tricks like this did not endear hauliers to the general motorists and the railways, too, were more than upset about the erosion of their traffic by road-going vehicles. Their pressures prompted several reports on transportation mainly chaired by people who felt that the motor vehicle had destroyed all the beauty and peace of life.

The ensuing legislation created the framework of haulage laws which was to last, in modified terms, right up to the modern day. Most remembered was the creation of the dreaded A, B and C operators' licences, but an even more important landmark in the Pickford destiny was when the four main railway companies were allowed to purchase equal quarter shares of Hay's Wharf Cartage Company. This, of course, meant that amongst other things, these railway companies were now the owners of Pickfords and Norman E. Box.

Even with stronger financial backing, MRS refused to be overawed by big Box and in an almost classic move they ensured the disappearance of this famous company. Whilst explaining what MRS did, it should be explained why they did it. Box had established their business so well that just their name implied a heavy haulier of great repute. Quietly at work in the Liverpool area was one Edward Box, not directly a competitor, but he did own a traction engine and his name was recorded as a business title at Companies House so, when MRS bought that business out, they were quite entitled to use that name. Hay's Wharf were furious that MRS had virtually stolen the use of their heavy haulage trademark and took them to court over it. But, even with the best of lawyers, this was one case they lost and MRS now had legal backing for their name of Edward Box Ltd. As the names were too similar, Hay's had very little in the way of options open to them, so with great reluctance the Norman E title was dropped and the vehicles repainted in blue and white. Pickfords may have gained a fleet of heavyweights, but heavy haulage had lost a company of great pedigree.

Even in 1920 the most efficient way to move tank rail locomotives, especially in South Wales, was to use Wynns Heavy Haulage who operated both light and heavyweight versions of the Fowler steamers.

Right and below right Norman E. Box's *Atlas* ending a haul to Bolton in the mid-1920s and needing full use of its power winch to get the BTH transformer to its final unloading position.

Below Surely a very early version of the adjustable trailer seen about 1923 on the quayside at Newcastle about to unload this Clarke Chapman vessel bound for Northern Ireland.

Left and below left The new articulated machinery carriers produced by Scammell were a big advance in heavy haulage, offering the luxuries of an enclosed cab and positive steering. They were soon brought into service by people like Rudds of London and even Norman E. Box utilised them for the smaller loads.

Left There was little doubt about what this Box outfit was hauling from Bury to Bristol about 1928, the 60-ton cylinder being an ideal fit for the Fowler rubber-tyred trailer. *Atlas* and *Talisman* seem to be losing a bit of water as they prepare to trundle off down to the West Country at a steady 15 miles per day.

Right R.T. Wynn standing beside one of the Welsh fleet's first articulated Scammells. **Below right** The company still kept their options open by operating this Foden steamer of similar configuration.

Right Showing the innovative thinking of Norman E. Box, this Mammoth Major was one of only three eight-wheelers built by AEC in this manner. Seen in the works of C.A. Parsons in 1936, the 15-ton transformer was well within the capacity of the low-slung vehicle.

Above One of the two big Scammell low-loaders operated by the general hauliers Curries is seen in the Heaton works of C.A. Parsons in 1938. The 54-ton stator was hauled to Newcastle quay prior to being shipped to the other Newcastle in New South Wales, Australia.

Left and below left December 15 1943 saw *Leaping Lena* about to leave BTH at Rugby with one of three 100-ton 242 KV transformers for export to the USSR. Even headed up by one of the ex-Box steamers, the winter cross-country journey to Manchester docks was recorded as having taken ten days to complete.

2. The Ts and Pacifics

As the Second World War approached, heavy haulage vehicles were very much in a period of stagnation. True the chain-driven Scammell articulated machinery carriers were not capable of 40-50-ton loads with their doddlers (little sprockets) on but above that the choice of vehicle was limited. Some of the steamers were now close to 40 years old and definitely overdue for replacement, but, apart from the two 100-ton artics, they were all that was left to pull the heavy loads. *Leaping Lena* was now hauling a Crane semi-trailer similar to that used on the Box outfit and both vehicles now sported a different engine. Owing to the foresight of Nottinghamshire bus operator Trevor Barton, the Patricroft diesel engine manufacturers of Gardners were now fitting their modified marine engines into trucks. The two big artics proved far better vehicles for the fitment but using them to haul 60- or 70-ton loads was mainly because there was not an internal combustion engined vehicle in existence which could pull them. Heavy hauliers became obliged to try the Scammell 'Coffee Pot' Pioneer which, although designed more for military, oilfield or colonial purposes, was more than strong enough to haul this sort of weight. Its role was that of an off-road machine so to the heavy haulage driver it gave the impression of a big, slow, numb motor. But in the darkness of hostilities one bright light was the arrival of a very good tractor, albeit from overseas.

Even the name Diamond T has a mystical sound about it and when you add a Hercules engine to its description, the picture of a thoroughbred is complete. The vehicle lived up to its image. It, too, was designed as a tank transporter but with 176 bhp and 12 speeds in the transmission line, it had lots more power and twice the gears of its British counterpart. Only rated at 85 tons gtw, Pickfords quickly appreciated that two Ts in pull-and-push combination were man enough to haul the heaviest loads of the day. Weights were building up with heavy electrical equipment out of places like BTH at Rugby, English Electric at Stafford, Metro Vickers in Manchester and C.A. Parsons in Newcastle being regular traffic. Moving stators, and transformers up to 100 tons in weight from inland towns was not without problems. Cranes' modern low-loading trailers were strong enough but the axle weights on the solid rubber tyres were anything between 30 and 40 tons per line. The 100-ton artics also ran with massive loadings, the consequence being that at times they literally sank through the road surface. The driver could feel this by a slowing down of the outfit and his first option was to try to accelerate out of the soft patch. But even with the new high-powered modern American trucks, putting your foot down with an all-up weight approaching 200 tons did not produce a great deal of effect. The remaining option with a sunken outfit was to jack it out and then plate the road with steel sheets. The crews were kept on their toes for, once progress was recommenced, they did not want to stop again so the recovery of the equipment was done at a very sharp rate.

The highways authorities were livid about the destruction of road surfaces and large bills regularly arrived in Pickfords' mail. They were not all necessarily paid as heavy hauliers became experts in the examination of road surfaces. If the material used was at all sub-standard then the cost of the damage was contested with never-ending arguments back and forth. Whose fault the holes were was a matter of conjecture but, in 1945, the consequences of this lack of strength were shown to be quite frightening. To the crew of the Diamond T pull-push outfit, taking an 80-ton roll housing from Sheffield to Falkirk was a straightforward haul. The heavy load route was followed north and sleepy Boroughbridge encountered as a matter of routine. Crossing the River Ure had been done on countless occasions but this time the trailer sank whilst half-way across. Wally Scott, the driver of the leading T, quickly realised that this was no time to reach for the jacks and plates and told everybody to bale out and run. Hardly had the men scampered to safety when the bridge collapsed throwing one of the Ts, the new Crane trailer and, of course, the roll housing into the river. To put it mildly there was hell on. Britain was endeavouring to get over the war and losing one of your main north-south highways does nothing to aid transportation or communications. On top of everything Pickfords had to recover the pieces which, in itself, was a major operation. The Army offered their help, which was gratefully accepted and they in turn moved the river, or strictly speaking altered its course so that the part of the bed where the debris lay remained fairly dry. It was then the strength of winches which dragged the pieces back along the river and up the bank prior to disposal. The tractor and trailer were complete write-offs whilst the roll housing was wiped clean and found to be undamaged which was, at least, one item less to pay for.

The end of the war saw the end of the steamer; not submitting to the modern diesel tractor but definitely succumbing to old age. They had served Pickfords well, one of their finest hauls being as late as 1942. A massive 120-ton casting was hauled from Sheffield to Distington Engineering at Workington straight over the top via Woodhead. Four of the locomotives were required and at 12-15 miles per day it seemed to take an eternity but the load was delivered. The choice of replacement was still limited and although Scammell had revamped the Pioneer calling it an 80-ton tractor, it was still slow. The T was by no means perfect either, a claustrophobic cab, headlights far too small and a tendency to snap halfshafts if the power was poured on too hard, but it could pull. Rudds of London favoured the big Scammell-Crane outfits as did Issac Barries of Glasgow. Fellow Glaswegians, Youngs, had a rare Diamond T artic on their fleet but the biggest thorn which was to stick in Pickfords' side was to come from South Wales.

Wynns had been hauling since 1863 and 80 years later had built up such a big fleet that they could virtually tackle anything. As early as 1890 they had been able to offer a heavy haulage trailer for hire with a capacity for massive 40-ton loads, but by 1943 double that weight was fairly regular. The company was trying hard to get into the fast-developing electrical industry which was producing some headaches amongst its customers. The big Crane trailers could carry the weight but the running height especially with transformers was far too high. Out of the necessity to drop the load as low as possible came the concept of carrying it slung on lugs resting between two girders. These in turn were located at either end on bogies so the birth of the girder trailer was seen. For the bigger jobs the load itself was hung between the fore and aft bogies and the first end suspension move was created. An Edward Box outfit is recorded as being the first combination to utilise this latter method of transportation hauling a heavy Metropolitan Vickers transformer from Manchester to Portobello, Edinburgh, in 1948. Science, as opposed to sheer brawn, was giving the heavy haulier a lot more power to his elbow but no amount of strength could withstand the sledge hammer which was dropped on the industry in that year.

The merits of nationalisation are best discussed by politicians but the end of the 1940s and the beginning of the 1950s saw a total upheaval in the industry with vehicles being bought and sold, repainted and redistributed back round the country. The Act of Parliament declaring government ownership specifically excluded compulsory purchase of vehicles used for abnormal loads, house removals, tankers, timber and meat haulage so, in theory, the heavy side of haulage should have been left alone. In practice it was destroyed. Pickfords, for instance, were owned by the four main railway companies and as this was the first method of transportation to be bought up it meant that they, too, became an asset of the country. Other hauliers also found themselves bought out because of 'other' reasons and some found most of their fleet compulsorily purchased whilst the odd heavy tractor they owned was excluded. The end result was that nearly every old company in the heavy haulage field just disappeared into the British Road Services machine although thankfully, for historians at least, two big names survived the major surgery.

The newly created BRS was split up into eight territorial divisions spread countrywide but owing to the influx of vehicles which were not really catered for in general haulage, a ninth division with its head office at Enfield, North London, was set up. Bureaucracy may have had its rules, but government officials did realise that you cannot just wipe out 300 years of tradition, so the division catering for house removals, meat, tankers and heavy haulage was, as we all know, called Pickfords. Twenty-six depots for low-loader work alone reflects the size of this specialist group but, in fairness, only a few of these catered for the very heavy traffic. Birmingham had always been one of Pickfords strongest depots along with Manchester and Sheffield, but come the revolution, Newcastle and Glasgow also slipped into this ultra-heavy league.

The transition for the men was a difficult time. One day you have your own boss and pride in your own company, the next you are just a small cog in the massive firm of Pickfords, up until then your biggest arch rival. Ill feeling got to such a pitch with the men from Edward Box and E.W. Rudd that, in order to appease them, the names of these companies were left on their vehicles, although the colour was changed to the Pickfords' livery and the lion on the door was BRS. Even the original Pickfords staff did not want all this extra manpower, extra depots and extra assorted machinery to tarnish their image so all in all nationalisation was not a joyous time, except perhaps for the Welshman. Having lost all their normal vehicles to the government, Wynns threw all their energies into abnormal load carriage investing heavily in ex-American tank transporters to do all the hard work. Twenty Hercules-powered Ts were the mainstay, but the most interesting purchase was six tractors manufactured by the Pacific Car and Foundry Company. The mighty Pacifics proved an impressive sight on the highways of the land but before they got to work, Wynns engineers had virtually rebuilt them to their own design. Being powered originally by a 240 bhp Hall-Scott petrol engine, the M26 tractor only achieved 1 mpg, so in the first instance the Hercules was fitted as an alternative power pack. The transmission was interesting in that the 6 × 6 configuration consisted of a normal shaft-driven front axle whilst the rear axles were chain-driven. Strengthened sprockets and crown wheels were fitted, along with a specially designed wooden framed crew cab and, as and when the vehicles were completed, they took to the road. The finishing touch by John Wynn, in true Box tradition, was the affixing of names: *Dreadnought, Conquerer, Valiant, Challenger, Helpmate* and *Enterprise*.

Coinciding with their arrival, Cranes built two special girder trailers to carry 140 tons or thereabouts, one for Wynns and one for the Manchester depot of Pickfords. What was mainly different was that the tyres fitted were at last pneumatics, the 16 huge 16.00 × 20 24-ply rating Goodyears ensuring a smoother ride for the load and a cushioned effect for the road surface. In 1952 Wynns threw down a gauntlet to the ninth division when *Dreadnought* and a pushing T carried the heaviest piece of electrical equipment ever moved by road. Using the big Crane trailer, 150 tons of BTH transformer, the first of six similar loads, was hauled the 70 miles from Rugby to Staythorpe in Nottinghamshire at an average speed of 3-4 mph.

Pickfords appreciated they had a challenge on their hands and their inherited old Pioneers were rather overawed by the impressive Pacifics. Thankfully for them, Scammell came to the rescue with a brand new design which was soon brought into service. The Constructor was of 6 × 6 drive and like its new 4 × 4 partner, the Mountaineer, it would be difficult to describe it as beautiful. It was not built specifically to look good, it was built for heavy haulage and the 12.17-litre C6NFL Rolls-Royce engine pushing out 184 bhp could certainly do that. Coupled to the six-speed gearbox was a two-speed transfer box so the 12 ratios gave the driver quite a turn of speed as long as he could stand the racket inside the cab. The big Constructor's arrival saw the retirement of *Leaping Lena* and KD 9168, both thankfully preserved for posterity. Box's 100-tonner had gone to Pickfords' Glasgow depot on nationalisation and there had seemed no limit as to the weight these two old vehicles could carry, the only constraints being the strength of the road surface they were travelling on. *Lena* was once recorded as pulling 180 tons internally in a Sheffield steel works and this was with the six-cylinder Gardner engine as well. Although the tractors were withdrawn owing to old age, their two semi-trailers fitted with suitable dollies continued to be used as drawbar outfits well into the 1960s reflecting, indeed, how well these vehicles stood up to the rigours of time.

Pickfords and Wynns were the leading heavy hauliers at this time but with the coming of denationalisation the 1950s saw a big growth in the competition, especially in the lighter end of the field. The London area had Annis and Co and Hallet Silbermans with Hopes at Bedfont, whilst Hills of Botley serviced the south coast. The midlands had Wrekin Roadways, Red House Motor Services and Starr Roadways, whilst Lancashire had Becks of Stockport, Walter Denton and Millers of Preston. Pickfords' Glasgow depot had strong competition growing right on their doorstep in the form of Gavin Wilkies, Glasgow Hiring and McKelvies of Paisley whilst the North-East of England was to spawn three major heavy hauliers which certainly gave Pickford's Birtley and Stockton depots a very hard time. In lofty north-west Durham, Siddle C. Cook's fleet grew hand in hand with the increased demand for Consett Iron Company steel and the number of licences he could get from the local traffic commissioners. Similar constraints were placed on the brothers Sunter, who were endeavouring to serve the Teeside steel works from rural Northallerton whilst midway between was a haulier just interested in weight. Crook and Willington Carriers were based at Bishop Auckland so, in theory, they were well placed for Consett, Tyneside and Teeside, an ideal location for their small fleet of Diamond Ts, Scammells and Fodens.

Cooks and Sunters had been competitors in both the timber and long steel traffic for some time and their progress into heavy haulage coincided as well.

Both started off with Mountaineers, RUP 900 and KVN 604, then in 1955 both stepped up to the big Constructors, SPT 600 and NAJ 920. Siddle Cook paid out £7,300 for his big Scammell and it was probably the only UK 6 × 6 Constructor ever to operate as an artic. Cook abhorred the waste of having to carry dead ballast around on his tractors and used the artic configuration so that the weight of the load maintained traction for the driving wheels. Sunters' Constructor reached an exalted place by being road tested by *Commercial Motor* magazine prior to delivery in 1955. John F. Moon recorded that he got 2.63 mpg at 85 tons gtw from the Rolls-Royce engine and it should be noted that the specification on the vehicle showed only Clayton Dewandre air servo brakes on a vehicle which could do nearly 40 mph and was to pull over 200 tons.

Compared to the hard-up private enterprise operator, who had to make do with adapting strong tank-carrying trailers for his heavy loads, Pickfords were able to afford the capital to invest in the biggest and the best. Working to their specifications, Cranes of Dereham produced TM 413, six axles of girder trailer with a capacity of 200 tons. What a monster! It was only one trailer but it meant that Pickfords could carry anything that heavy giving the spread of weight required on a girder trailer. The solid-tyred bogies offered this type of capacity but were no good for concentrated lumps like transformers or roll housings. Wynns retaliated and immediately applied for a licence for a similar 200-tonner and, of course, Pickfords objected saying they could meet all the demand made to move loads of this weight. Fortunately for Wynns both AEI and BTH, two of the biggest electrical manufacturers, supported their application, as they did not want BRS to monopolise the traffic, and the licence was granted.

If anyone held a monopoly it was Scammell whose vehicles seemed to dominate the heavy haulage scene at the time. Their articulated eight-wheelers were the standard for 20-25-ton loads, then the Mountaineer pulling a drawbar trailer came in at 45 tons. The Constructor was good for 100-plus and in 1956 Scammells bridged the gap between these two by introducing a Leyland-engined 6 × 4 tractor. The Junior Constructor was one vehicle which could be described as almost beautiful, although in modern day terms the 150 bhp from the 680 engine would be thought of as vastly under-powered. Pickfords did not think so and, like several other operators, got a very good work load out of their numerous Juniors.

Wynns may have been impressed but were not tempted to change horses and persevered with their American contingent for even if it meant double or triple heading, the red Pacifics could match the blue Constructors. Both the companies offered equal capacity 200-ton trailers now, but Wynns were able to take a leap on the ninth division with the arrival into the country of the first flat top. Late 1957 saw a 160-ton capacity, eight-axled, 64-pneumatic-tyred trailer go into service made by Willy Scheuerle Fahrzeugfabrik of Germany. Quite a mouthful, but quite a trailer. In appearance it had similarities to Pickfords' old solid tyred bogies joined together with the 10-ton table. Both were automatically steered but where the Scheuerle more than scored was with its hydraulic suspension. At the touch of a button the running height was raised/lowered from 2 ft 4 ins to 3 ft 6 ins so loading and unloading was now a matter of seconds as opposed to the previous hours needed for jacking.

The gangs of Pickfords and Wynns were renowned for moving objects far and beyond the capacity of the heaviest mobile cranes, which were limited then to 40 to 50 tons, by winching, jacking, skidding and also adopting techniques that could not be found in any text books. The arrival of this new-fangled trailer did not change the requirement for this art but it certainly made things easier for the opposition. If this was not bad enough there was yet another competitor waiting in the wings to show how he, too, could out-manoeuvre the experts in a way never quite attempted before.

Left Woodhead 1942, Pickfords big Crane trailer and four of the big steamers in their finest hour, en route from Sheffield to Workington with this 120-ton casting.

Right This all Fowler outfit is seen at Rugby on October 18 1942 proving that these ex-Box vehicles certainly worked for their keep.

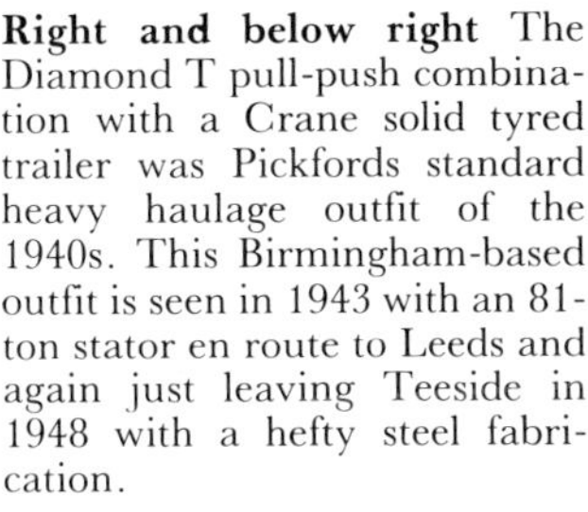

Right and below right The Diamond T pull-push combination with a Crane solid tyred trailer was Pickfords standard heavy haulage outfit of the 1940s. This Birmingham-based outfit is seen in 1943 with an 81-ton stator en route to Leeds and again just leaving Teeside in 1948 with a hefty steel fabrication.

NO ROAD
SLOW

Above As early as 1943 Wynns were using this girder trailer having a capacity of 90 tons. It is seen in 1948 leaving BTH at Rugby headed up by a very special Foden rated as a 100-tonner.

Left Boroughbridge 1945. The devastation caused to the bridge, trailer and Diamond T is plain to see, whilst the No Road sign seems to say it all.

Below *Leaping Lena* and driver Lowther, in the white overalls, about to leave C.A. Parsons in 1948 with a 124-ton stator for delivery to Newcastle quayside prior to being shipped to Canada.

Left and below left Two of the numerous loads hauled out of BTH Rugby bound for the Clunie power station in northern Scotland late in 1948. These pictures show the differing types of outfit used by Pickfords around that time. The old chain-drive Scammell is obviously happy on pneumatics hauling a 23-ton half stator frame, whilst the Diamond T is slightly 'over spec'd' with this awkward rotor rim weighing in at 30 tons.

Left Well thought out rigging of a set of girders and solid tyred bogies were made to ease the carriage of this BTH transformer on November 15 1943. Weighing in at 71 tons, it would be no problem for the Diamond T-Scammell box tractor combination on the short cross-country journey to East Kirkby, Nottinghamshire.

The first recorded end suspension move was this Metropolitan Vickers transformer which was hauled from Manchester to Portobello, Edinburgh, in 1948. The outfit is Edward Box's but the BRS sign on the side indicates that they, too, had already been swallowed up into the government machine.

At 110 tonnes this locomotive, destined for India, was the largest single exhibit in the Festival of Britain held at Battersea. The haul from Surry commercial dock to site in 1951 was the first outing of a Wynns Pacific and the company was extremely pleased with the way in which Fairfields of Chepstow had produced these special supporting beams at very short notice.

The Scammell tractor with four in line semi-trailer was virtually the standard 20-25-ton low-loader until well into the 1960s. This outfit, part of the Edward Box fleet prior to nationalisation, is about to leave Workington in 1952 and shows great faith in its winch rope.

Above left Jack Henderson checks security prior to leaving Stockton with this vessel bound for Birkenhead in 1953. There was no weight trouble for the 45-ton box tractor—the biggest problem was how to hang on to the attendant who wanted to bale out every time the outfit drove under some high voltage cables.

Above Seen in BRS (Pickfords) colours, this little box tractor belonged to the fleet of E.W. Rudd prior to nationalisation. The 37-ton Ashmore vessel had a long, bumpy ride ahead of it on these solid tyred bogies from the Stockton works to a BP refinery in Kent.

Left Birtley-based Norman Eaton drives down Northumberland Street in Newcastle on November 12 1953 with a 46¼-ton locomotive out of Robert Stephenson's Forth Banks works bound for Birkenhead. This immaculate Scammell, having a top speed of 16 mph, eventually finished its life working hard for Sunters of Northallerton.

Left Sunters first heavy haulage outfit, bought new after denationalisation, was this Mountaineer-Crane float combination. One of its first jobs was to carry this RB machine from Boreham Wood to Grangemouth, but driver Robinson had to take a circuitous route owing to height and weight limits. Every steep hill saw the little Meadows-engined Scammell run out of breath and a double head had to be begged from anyone passing. North of Thirsk the ex-BRS Foden in the background followed behind and hooked up when required. That arrangement worked for the rest of the way apart from at Coldstream, where a third vehicle was needed to mount a steep incline.

Above and right Master Raymond Cook sits on the flagship of his father's fleet on the day it was collected new from the Scammell factory in 1955. SPT 600 was probably the only Constructor ever to operate as an artic, a configuration prompted to cut down the necessity of having to carry 15-20 tons of dead ballast weights.

Above The DW registration might suggest that these Mighty Antars were part of the Robert Wynn fleet in camouflage. In fact they were on test by Wynns to demonstrate to their Australian owners that they were more than capable of coping with whatever the Snowy Mountain Range may demand from them.

Left Described as being the first application in the UK of a trailer manufactured by Willy Scheuerle Fahrzeugfabrik, this Wynns 32-wheeler had a capacity of 150-tons and heralded the start of flat-top hydraulically suspended load carriers. It is seen leaving Ipswich with the 77-ton base of an excavator en route to Heddon-on-the-Wall in Northumberland.

August 31 1955 sees this Manchester-based outfit on its two-day journey down the Tyne valley from Heaton to Swalwell with a 127-ton Parsons transformer destined for the Stella West power station. One of the 80-ton Scammell tractors broke down en route but the remaining two were more than enough to finish the short haul on their own.

A mixture of three gleaming Scammells wait on December 1 1956 whilst the Birtley crew jack down the Crane solid tyred trailer before squeezing under Walker Bridge on Tyneside. The 130-ton Parsons transformer was en route to Newcastle quay to be shipped to Liverpool.

May 20 1958 saw this mixed Birtley-based outfit leave the Heaton works of C.A. Parsons with a 78-ton turbine shaft destined for Canada. The smart Junior is being pushed by an old 80-tonner whilst the carrying is being done by a Crane trailer which was known by all who worked on it as 'the abortion'.

These pages Described by the media as being the largest load ever moved by road at that time, this 131 ft long, 16 ft diameter, 93 tons of stainless steel took three days to be hauled the 18 miles from Ashmores' Stockton factory to ICI Wilton in January 1957. Andy Higgins driving DUS 951, an ex-Issac Barrie Scammell, and the Crane solid tyred bogies from Pickfords Glasgow depot, managed most of the way unassisted. To surmount the 1 in 7 Leven Bank, two Birtley based 6 × 4s joined up; although the time taken to travel the 150-yard incline was 13 minutes. There was a tense moment at the steepest part of the climb as the 50 ft rear overhang nearly grounded. The outfit was stopped and planks placed in front of the rear bogie so that the load would be lifted when it moved off. The leading tractor had slight wheel spin as it restarted but the hardest part of the journey was eventually finished and the load delivered damage-free.

PICKFORDS

PICKFORDS
PICKFORDS

WYNNS
NEWPORT · CARDIFF · LONDON
DREADNOUGHT
192
192
GDW-277
WYNN'S

WYNNS
POWER TRANSFORMER
WYNN'S
WYNN'S
HEAVY HAULAGE
NEWPORT · CARDIFF · LONDON
192
192
GDW-277

WYNNS
NEWPORT · CARDIFF · LONDON
DREADNOUGHT
192
192
GDW-277
RUGBY

These pages The Wynns transformer trains of the 1950s were a regular sight in Rugby. *Dreadnought* and friends are en route to Staythorpe in Nottinghamshire with a BTH unit weighing in at a mere 140 tons, easy stuff for the six-axled Crane trailer. The 176 bhp Hercules engine did prove rather under-powered for these loads, thus requiring the use of three tractors although it was not surprising as the all up weight was close to 300 tons.

Above Ex-tank transporters were greatly utilised by most heavy hauliers not least Cooks of Consett. Siddle C. Cook also operated buses, an example of which is seen in the background. Close co-operation with Consett Iron Company meant free use of the occasional casting for ballasting purposes only.

Left KVN 604 is seen working hard in this scene from 1958. The 100 ft long pipe was destined for the number three Olefine Plant at ICI Wilton. The trailers used were ex-Army transporters.

Right Not a great deal of weight but still very awkward, this flame deflector is seen on the Roman Road at Greenhead en route from Ashmores at Stockton to the rocket testing site at Spadeadam, early 1958. A novel use is made of corrugated sheets on Sunters Foden which was fitted with a Gardner engine and eight-speed gearbox.

JPY 837

Above In 1959 this 30 ft high, 21 ft wide, 130-ton boiler was the largest yet to be made by Vickers-Armstrong at Barrow. It is seen in the capable hands of Wynns en route to the Buccleuth dock to be fitted in the *S.S. Oriana*.

Left The Rotinoff is seen leaving Hawthorns works at Darlington in mid-summer 1959 with the 150-ton capacity Crane trailer. The locomotive destined for export from Liverpool docks to India was similar to the load hauled by the Super Constructor when it was demonstrated by Scammells to Sunters at the end of the 1950s.

3. Bradwell and the Rotinoff

With the death of Joe Elliott, the vehicles of his company, Crook and Willington Carriers, came up for sale and as licences for heavyweights were like gold, they were quickly snapped up. Siddle Cook bought one of the Fodens, but Sunters picked up the other five, Ts, Fodens and a Scammell. This certainly strengthened the Northallerton fleet but the springboard into some real heavy business came along when the construction of Bradwell power station in distant Essex was being planned in 1956 and Tom Sunter was not the sort of man to turn down an opportunity like this. Head Wrightsons of Thornaby had been asked to tender for the making of 12 enormous heat exchangers for installation in the heart of the power station, the hitch in the contract was that if Heads agreed to build them, then Heads were responsible for the delivery of them. Even with modern day sophisticated equipment, a road haul of 250 miles of a vessel weighing 235 tons, 92 ft long and 22 ft in diameter 12 times over was out of the question, so transportation was to be quite a problem. Head Wrightsons were famous for their dock gates and the method of delivery used for these was simply to launch them into the River Tees then tow them by tug to wherever they were required. So models were made, tests carried out and when it was found these boilers could also be floated in water without causing any internal damage, the tender to build was accepted and work commenced. The only remaining problem was how to move them the 1½ miles on dry land once they had reached Essex and it was at this point that the brothers Sunter came into the picture.

BRS (Pickfords) Ltd obviously had both the equipment and the expertise to tackle such a job as this, but the manufacturer obviously appreciated being able to talk to the man at the top rather than speak to a manager, who would talk to an engineer, who would report to his superior, etc, etc. Sunters did not have the equipment at that time to do a job like this nor had they tackled anything like it before, so it must speak mountains of the personality of Tom Sunter for him to be able to convince Heads that his team could do the hauls, but he did and the job came to Northallerton.

Sunters went to Cranes of Norfolk to supply the trailers for the job, two solid-tyred bogies each rated to carry 100 tons. At 10 ft 6 ins they were slightly wider than normal but it still meant that through each of the 16 little wheels per bogie there would be 8 tons transmitted to the road surface. The tractor unit was more of a problem as plans of the move showed that the boilers would have to be pulled out of the water on an incline of 1 in 10. Even with one tractor the all-up weight would be close to 300 tons so it obviously needed two or even three pulling to be on the safe side. Three Constructors would do, as would three Pacifics, even a combination of Sunters' tractors already in the fleet may well have sufficed, but on to the scene came a man who was trying to prove that his vehicles could do the job just as well as Scammell, Pacific or anyone else. Rotinoff might be a name remembered now, but in 1957 few people had even heard of him. His parents were Russian but he was born in this country and had grown up to be quite an engineer. Working from a small trading estate just off the A4 at Slough, he had designed and built on the lines of the Diamond T, a Rolls-Royce powered 6 × 4 ballasted tractor which he reckoned was the motor Sunters should have. The brothers must have agreed for RPY 767 soon came to shatter the tranquillity of North Yorkshire.

May 1958 saw the first heat exchanger leave Thornaby on its stopwatch-timed journey down the North Sea. The preparations on the River Blackwater included a special double wooden berth to position the boiler and a concrete strip to act as a slipway out of the water. When the tide went back, the two Crane bogies fitted with suitable cradles and connected by a central towing bar were reversed into position and anchored down. As the river rose and submerged the bogies, the waiting boiler was winched into the exact position alongside the berth. Held like this it gently sank, as the tide fell back again, right on to the two waiting bogies. When the tractors hooked up, the bogies were released from their anchorages and things were just about ready to go, although it did not go anywhere as the trailer brakes had seized. Even with lashings of protective grease, water had seeped into the lines making the system totally inoperative but, with no time to hang about, it was a matter of draining everything off and running without any brakes at all on the two bogies.

For the first boiler Sunters' own Atlantic was headed up by a further two from Mr Rotinoff's factory. Obviously a publicity stunt but if they did not do it this man's reputation was gone forever. Being prototypes all three tractors were different. Sunters' one had big single driving wheels, the other two had smaller twins. The rear axles had different ratios and the gearboxes were different. Undaunted, Mr Rotinoff had worked it all out and although a lot of thought and discussion had preceded the first pull it was still a worrying time. At the drop of Tom Sunter's handkerchief, the three Rolls-Royce engined Rotinoffs bellowed forth and, looking like something out of *Quatermass and the Pit*, the first boiler crept out of the river. Up the 1 in 10, the gradient eased to 1 in 46 then eventually levelled out. Here the two factory Rotinoffs were unhooked and John Robinson, with RPY 767, finished the haul alone at near to 300 tons gtw. It was a tight squeeze at times and a few trees were pulled down, also the odd stretch of road was marked as the Atlantic dug in to give itself traction.

Over the next 18 months the remaining 11 heat exchangers were delivered in a similar manner without the slightest of hitches. A varying combination of tri-heading tractors was used to move the boiler out of the river but normally they were the Rotinoff, the Constructor and an ex-Pickfords Pioneer, MUA 461. It reflects on the gearing of the old Scammell that, whilst the Atlantic and the big 6 × 6 both started off in 'crawler', it moved off in second. Much praise must be heaped on drivers such as Robinson, Fraser and Emms who, at the drop of a handkerchief, moved 370 tons dead weight on a 1 in 10 incline, using tractors with manual gearboxes and clutches, yet without a suspicion of transmission failure. In 1982 this sort of weight is fairly usual but 24 years earlier it was heavy stuff. John Robinson and the Rotinoff pulled all 12 boilers, except one. Running into Merseyside one weekend his instructions were to drop the loaded trailer and run down to Bradwell solo. He had only gone 14 miles when with an almighty bang, the reduction gearbox seized up snapping the prop shaft and half the transmission with it. It was up to Jock Fraser in NAJ 920 to do the necessary with this boiler—300 tons with a Constructor is good going in anyone's books.

The transmission of the Rotinoff proved to be the weakest part of the vehicle. A six-speed main gearbox transmitted via a three-speed transfer box and the spread of 18 ratios gave the vehicle a capability of 48 mph. Sunters worked the vehicle hard as a lot was expected of it but it just could not take the strain. The tow bar proved to be John Robinson's best friend as he was barred back to the garage from all over the country. The teething troubles in the vehicle's first five years might have broken any other company, although Sunters hung on to it. In 1962 the vehicle was transformed with the fitting of a Self-Changing Gears eight-speed semi-automatic gearbox and went on for a further 14 years of very productive life. The Atlantic now graces the Transport Museum at Swindon, it being one of the few Rotinoffs left in existence. With the death of the founder no one fully took over his business, although Atkinsons eventually became the spares supplier and it may be worth comparing the lines of their ill-fated Omega tractors with that of the Atlantic. As it was the prototypes were left to stand up for themselves and Rotinoff just became a part of history.

Pickfords were probably the only heavy haulier which could have come up with a sufficient order to continue production of the Atlantic but, being something of an unknown quantity, BRS opted for the proven Scammell range the latest model being the Super Constructor. WYH 901 and 902 were two early examples being based at Birmingham and in concert with the Crane 200-ton trailer were a very impressive sight and fit to haul the heaviest of concentrated loads moved at that time. The Supers were a big improvement on their old 6 × 6 stablemate, 901 and 902 having the 15-litre Leyland-Albion 900 engine. The supercharged Rolls-Royce was an option but the biggest step up was the removal of the clutch pedal and the fitting of the semi-automatic gearbox. Gavin Wilkies were obviously impressed and a gleaming example in blue, 473 AGG, soon came to grace the Glasgow fleet hauling their solid-tyred bogies throughout the country. With Wynns being inseparable from their American trucks, now using the Meadows engine, Scammells tried to encourage Sunters to purchase one of these new vehicles and asked if they could demonstrate their top of the range model, so a trailer and load were laid on. A lot of locomotives were being hauled both from Darlington and Teeside about this time, so one of these across to

Liverpool would suit both Scammell and Sunters. A brand new demonstrator came along with the best of Scammell's drivers and, although the weathered Rotinoff followed behind, the Watford men were out to prove their 6 × 6 was a better tool than the Atlantic. Any low-loader driver who had to circumnavigate Leeds on the heavy load route lived in fear of Rodley. The bank was not very long, but it was steep and right at the top you had to turn sharply. With artics the front wheels tended to come off the ground whilst with tractors it was impossible to get the necessary grip. The stone walls on the ring road were regularly robbed as drivers worried about ballast and traction but the Scammell pushed on undaunted. Sunters offered to hook the Rotinoff on behind but the Watford men, not knowing Rodley, felt this was an insult and tackled it alone. In fact it nearly made it, but close to the top traction was lost and in desperation the Rotinoff was asked to push. All that did was to dig a big hole in the ground as it fought for grip, for to stop on Rodley is a recipe for disaster. It took the aid of an old bonneted Leyland breakdown truck to head up before the bank was cleared, but this incident apart, the Super was a good truck and soon two of the Rolls-Royce version came into the Sunter fleet.

The Northallerton operator still failed to worry the big two who continued to try and out punch each other with bigger and better equipment. In May 1962 Pickfords introduced their 230-ton capacity ten-axled Crane girder trailer heralded as the most sophisticated yet, although the specification still showed that it had hydraulically-operated Girling brakes. Vital dimensions were: 89 ft 11 ins long, width variable between 9 ft 10 ins and 17 ft 4 ins depending on the load's requirements, plus an ability to drop its third and eighth axles if desired. In the first four weeks of its life it carried four transformers weighing in at 150, 155, 160 and 190 tons which indicates the work which was available to those who could do it.

In 1963 Wynns hit back with 'Britain's biggest trailer' again a Crane but this time on 12 axles with relatively small tyres and a capacity of 300 tonnes. Wynns' engineers felt they knew a thing or two about getting the best out of girder trailers and this one was made with the swan necks hinged on the bolster, an entirely new concept used for making easier adjustment to suit loads of varying width. The bogies were also detachable and great use was made of these when the big girders were not required.

Wynns' tractors had also taken on a new look, for with old age and rust striking hard on the Diamond T cabs they were replaced by a slightly bigger unit which was of different shape yet of characteristic appearance. New Cummins engines were also being fitted as the Meadows seemed prone to piston trouble, with the NH220 in the T and the NHRS 285 in the Pacific. Reinforcing the modernisation the Pacifics were fitted with the SCG RV30 eight-speed semi-automatic box so that there was now very little to choose between the Pacific and the Super Constructor. In the smaller tractor field there was now a new option aimed at ousting the likes of Foden and AEC for 30-40-ton loads. True, Wynns were selling Guys at this time but the Invincible was well liked by many others operating in this weight bracket. Both four- and six-wheeled versions were made, the Gardner 6LX 150 bhp engine being the standard power pack. Both Sunters and Cooks had several of these and both had a beefed-up version fitted with the 12.17 Rolls-Royce engine and ZF 12-speed gearbox, which was quite a powerhouse. About this time Cooks were getting good service from six left-hooker Super Beaver/Hippos and, although the drivers had to change gear with the wrong hand, they were a big improvement from the second-hand Scammells which were the backbone of the Cooks' fleet. Pickfords oddly did not go for the Invincible range, preferring the Scammell Highwayman and some special snout-nosed little 4 × 2 Atkinsons for their lighter work.

However, heavy haulage is not just about tractors and trailers, engines and gearboxes, it is about men coping with loads and some of the tales which are told are frighteningly funny. Lol Johnson was driving Sunters' Constructor north on the A1 near Ferryhill with a high Kaldo vessel when he saw a large flash come from underneath the lorry, so he immediately stopped to investigate. What he did not know was that the power from a 60,000-volt overhead cable had arced down on to the load owing to the prevailing atmospheric conditions. An automatic cut-out stopped the flow of current but as Lol and the mates were examining the vehicle, the power automatically cut itself back in. It arced once more and, with a mighty bang, driver and mates were lifted off their feet and thrown into the hedgerow. When asked later what steps he took about an ensuing fire in one of the trailer tyres, Johnson allegedly said, 'Big ones. As fast and as far from the wagon as possible'.

Jimmy Golding and Jack Emms of Sunters must hold the record for taking the longest time to travel the 36 miles from Thornaby to Consett with the elapsed period being six weeks. However, it was owing to the atrocious weather conditions and not the men and machines which prompted all the delays. The snow drifts were as high as the leading Foden as the 98-ton load was mounting a steep incline near Tantobie when, without warning, the tow bar to the trailer suddenly snapped as a result of the cold and the big 100-tonner leapt away. Although Jimmy was pushing hard in his Diamond T, the Scheuerle trailer gradually slid backwards into a snowdrift accompanied by shouts of 'Chocks! Chocks!' from the Golding cab. Jimmy was known for his caution with big loads as he was well aware of the poor braking of the current outfits, but it took a certain amount of medicinal lubrication from a nearby hostlery before the crews could steel themselves to resume their haul to CIC.

For one particular big tractor to stand out wherever it went meant that it needed a personality and identity all of its own. *Leaping Lena* came into this bracket as did Box's 100-tonner but where there were numerous vehicles in the same fleet virtually identical then, to anyone other than the crews, they were all the same. Pickfords had countless Constructors and Wynns a posse of Pacifics, although 192, *Dreadnought*, stayed in the fleet for over 20 years and must be the most famous, but it is the one-offs which are remembered. Sunters' Rotinoff comes into this bracket as did the two big 6 × 4 tractors built by, and called, Annis. Arguably PPY 264 is also remembered by a lot for it, too, was very much a one-off. Fodens were well-known for their specials and this Sunter tractor could pull a genuine 100-ton load, provided you were not in a hurry. Its eight-cylinder Gardner engine drove through an eight-speed gearbox but where the big Fodens got their strength was through the use of hand-operated reduction hubs. These gave the ability to pull, but it also meant the performance of the vehicle was that of a slow, ponderous, deliberate motor.

Out of all the Sunter drivers only Jack Emms had the personality which could accept and love this old tractor and he tells one story of the vehicle which I found impossible to believe but he swears is true. The local tyre man put eight new tyres on the rear bogie one day and, as normal, used soap on the beading to ease fitting. Straight out on to the job, Jack got as far as Six Hills near Bawtry, when he noticed that he had been alongside one particular tree for a long time, even bearing in mind the slowness of his vehicle. The engine was running, the gearbox was engaged, but even though the road was not icy, the vehicle was making no progress. The crew dismounted to find all the rear wheels revolving, but the tyres were stationary on the roadway. The liquid soap had allowed the wheels to slip on the beading and although the valves had snapped, they had self-sealed so the tyres were still inflated. The inner tubes had to be physically spiked and burst before the tyres could be refitted and, of course, from then on soap was never used in the fitting of new tyres on the lumbering Foden.

This old girl was strong but even all her power would not pull the loads that were to be moved 20 years later. The two decades following 1963 saw heavy haulage dramatically change with the 'big two' being ousted as the leaders in the field.

Seen in North Road, Darlington this Junior Constructor, with another Scammell pushing, is making steady progress for Liverpool with this 120 ft long Whesshoe vessel.

Above Consett Iron Company were not heavy hauliers in their own right but this shot, taken in the late '50s, shows how heavy haulage brings out the adaptability of normal road hauling equipment. This 130 ft long pipe had to be moved on site so the normal billet carrying trailers were utilised, the aft one in reverse form. The AEC Matadors were specially designed for CIC after consultation with Pickfords' engineers. They sported a Kirkstall drive axle and their usual bread run was to the mills at Jarrow, keeping them fully supplied with Consett steel.

Below NAJ 920 is seen entering Northallerton High Street with this 120 ft steel tower destined for a refinery in Lancashire in the late 1950s. Owing to bad bends in Thirsk town centre any long loads southbound from Teeside, like this 55-ton Ashmore vessel, were forced to leave the A19 and travel through Northallerton. The extra long drawbar was another Sunter special intended to inflict hernias on all who lifted it. Jock Fraser at the helm was most upset by having to drive this load through Manchester at midnight on New Year's Eve, a time when all Scotsmen get the call!

Above The first Bradwell boiler is seen in position having been winched to the exact point directly above the solid tyred bogies. The three Rotinoff tractors are seen waiting for the tide to recede in the top left of shot.

Left Once the tide went back, the tractors hooked up. Divers checked that the boiler had settled into the exact position then, with securing chains attached, it was ready to go. Mr Rotinoff is seen in the white mac waiting to prove that his three Rolls-Royce-powered prototypes could move 370 tons starting on a 1 in 10 incline using manual gearboxes and clutches.

There were always plenty of people on hand to give driver Robinson advice but, knowing him, he would follow his own judgement. The police escort had more trouble coping with the crowds than any vehicular traffic.

Varying combinations of tractors were used to get the remaining 11 boilers back on dry land. Here MUA 461 is seen backing up to tri-head the Constructor and Rotinoff. Tom Sunter is seen in collar and tie with Ron Allinson, one of the Head Wrightson boffins who conceived this method of transportation, being on his right.

November 1959, as the shadows lengthen the Rotinoff and the twelfth Bradwell boiler are shown in true perspective en route to the power station for the last time.

SAFETY FIRST
DO NOT BOARD A MOVING BUS
NO ENTRY
KEEP LEFT
DAVY UNITED
PICKFORDS
PICKFORDS
SRW 359
WYH 901

These pages WYH 901 and 902, two gleaming Super Constructors close to the end of a five-day haul in April 1960 from Davy United, Sheffield, to Consett Iron Company with this 170-ton roll housing. For the notorious Jaw Blades bank and other steep hills in north-west Durham PUC 472, an elderly Birtley-based 6 × 6, headed up the outfit making an all up weight close to 360 tons. TM 413 was an example of the strongest Crane girder trailer at that time, having a capacity of 200 tons and an unladen weight of 80 tons.

With rust and old age striking hard on the cabs of Wynns' Diamond Ts, the fleet took on a new look as they were replaced with a bigger unit, yet still of characteristic appearance. This outfit is seen leaving Rugby in February 1959 with the new Scheuerle trailer and a small Scammell pusher.

Dreadnought and the Scheuerle trailer used in two-bogie form easing over the old Green crossing near Newport Castle in 1960 as John Wynn offers instructions. The gantry girder being hauled is en route to the Uskmouth B power station.

Sunters' Rotinoff and 100-ton Foden go round South Parade roundabout in Northallerton bound for Erith in Kent, mid-September 1960. Even though it was running close to 20 ft wide and 180 tons gross, the traffic clerk badly underestimated the power of the Atlantic for, when it arrived at the Metropolitan boundary for the pre-arranged escort, it was seven and a half days early.

Right and below right February 19 1961 sees Sunters old Mountaineer on a round-the-houses move from Workington to Marchon Products, Whitehaven, with a journey time of six hours. The Scammell-Scheuerle outfit was obviously over specified for this 9-ton load, but the automatic steered trailer did provide the ample support needed to lift the 24 ft wide drum high enough to miss the lowly street furniture.

Right If Sunters were to haul this same mould from Thornaby to Consett today, it would be carried a lot closer to the ground even though the home-made semi seems well up to the job. This 11.3-engined Mammoth Major was well liked by driver Philip Braithwaite who said it would take 30 tons anywhere.

Left Jock Fraser punches Sunters' maturing Constructor through Bowes village on the A66 en route to Consett with the Kaldo ladle carriage which took up excessively long residence in Carlisle owing to a difference of opinion with the local highways authority.

Below Ted Stokes rides shotgun on VPT 85, one of the two Crook and Willington Diamond Ts bought by Sunters in about 1957. It is seen in July 1962 headed up by another T with an Ashmore Benson vessel en route to ICI Severnsides at Bristol.

Siddle C. Cooks quickly snapped up export cancellations which occasionally came on offer, like this Super Hippo due originally to go to Brazil. Changing gear with the wrong hand was no hardship to the drivers who thought the vehicles a big improvement on the large fleet of secondhand Scammells which Cooks favoured.

One of the biggest tractors in the Edward Beck fleet was this crew-cabbed Foden with prominent grill, a normal sign of the fitment of a Cummins engine. It is seen about to leave the Distington Engineering works in Workington with this hefty fabrication in about 1962.

Diamond Ts were extremely popular as medium weight heavy haulage machines but this Elliott outfit is a very rare artic version. This York haulier always managed to use semis to good effect. The Lima cab section is en route to Rochester in 1962.

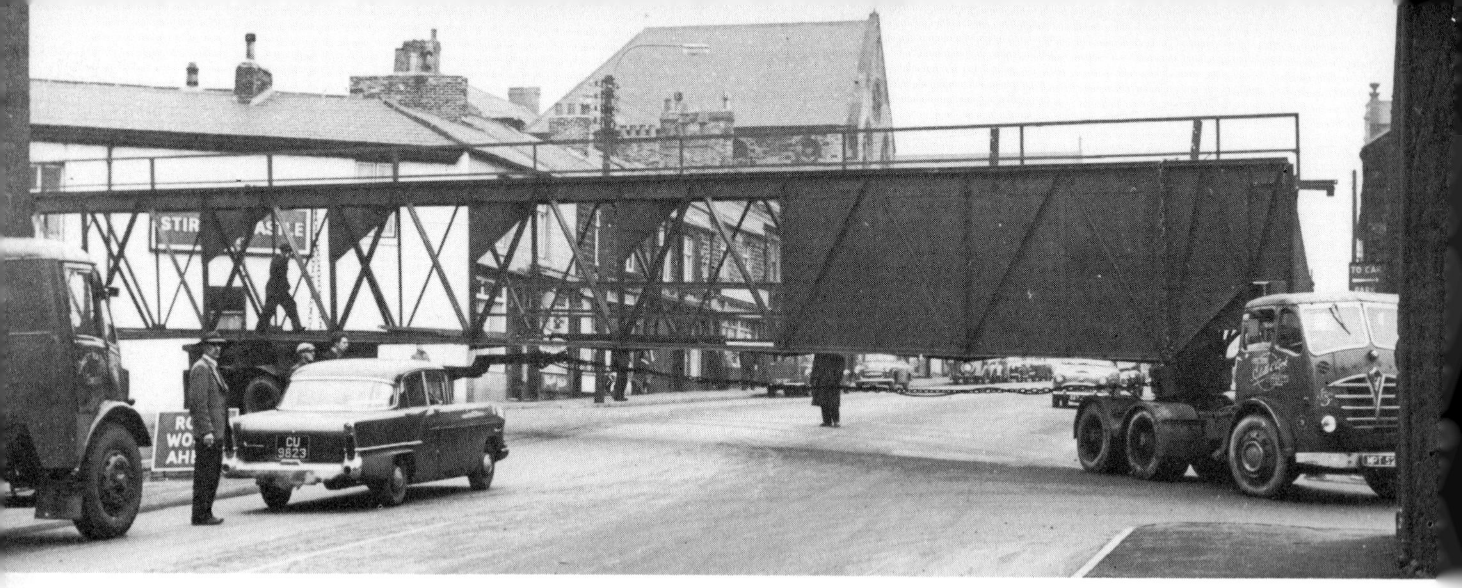
CU 9823

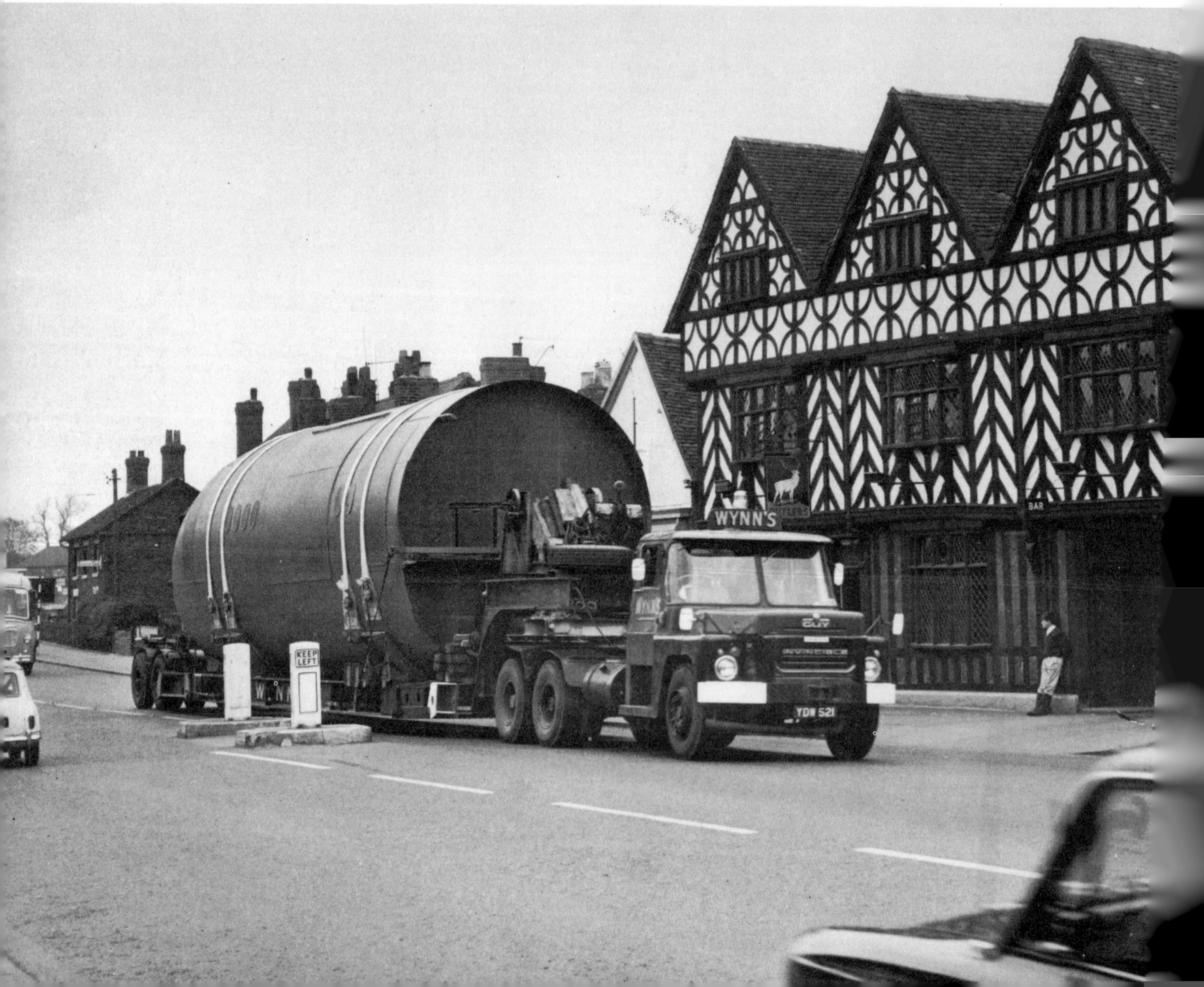
WYNN'S
KEEP LEFT
GUY
YDW 521
BAR

Left Siddle C. Cook seen operating this ex-Crook and Willington Carriers Foden with a chain-pulled pole trailer en route to the new CIC plate mill at Hownsgill. The Gardner 6LW-engined tractor with reduction box also had hand-operated reduction hubs and was said to have enough torque to climb mountains provided the prop shaft did not snap.

Below left Twenty special loads of VC10 fuselage sections, hauled from Preston to Weybridge by Wynns, required a specially extended Scheuerle semi-trailer and this load was hauled by a very special Invincible.

Top right *Conqueror* and *Dreadnought* in concert are an impressive sight as they make easy work of this Ferranti transformer. The small-wheeled girder trailer was the first Crane 300-tonner bought by Wynns in 1963.

Centre right This freshly painted Birtley-based Constructor is seen having a storage tank loaded on to its 85-ton capacity Crane girder trailer on September 18 1963. The BSC vessel was transported down the River Tees by barge which meant that the road haul into Teesport was kept to a minimum.

Right One of the many Guy Invincibles operated by Wynns is seen en route from Darlington Forge to Uskside Engineering, at Newport, with an awkward 50-ton section which required carriage on a Wynns-built tilt frame.

See page 53.

4. Changing times

In 1964 Wynns and Sunters joined forces. Both companies suffered family bereavements and with ensuing problems over death duties coupled with the necessity to have strong financial backing for their operations, both decided to sell out to the massive Bulwark United Transport group. It was the biggest fear of all the Northallerton staff that the managing director in distant Chepstow would feel that Boroughbridge Road would prove an ideal base for a north-eastern Wynns' depot but, as it was, both companies were allowed to operate separately, yet with great co-operation over exchange of information, machines and manpower.

Along with this build up of strength in their competitors, Pickfords were having their own internal difficulties. On the face of it, 26 depots spread countrywide for low-loader work should have meant a virtual monopoly of all the work on offer. With the biggest and best equipment made, no one should have been able to hold a candle to them but, as each branch was run as a separate entity and each was expected to run at a profit, they were on occasion in direct competition with each other. This came about as not every depot had identical equipment and only the big five or six could deal with ultra-heavy traffic. Managers of all branches fiercely guarded their own customers, so even though another branch may have been in a better position to deal with a particular load, they were loath to part with the move, preferring to sub-contract and get the 10 per cent. Occasionally greater use of outfits could have been made to cut down emtpy running as, once again, branches were loath to give up traffic to their 'rivals' for back loads when the originating branches' vehicles could have carried it in due course. But with the 1950s and '60s being a virtual boom time in heavy haulage it did not make a lot of difference as there was so much work that everyone was kept busy.

The managers also had other things on their mind. Having battles with head office to get more and better equipment was bad enough but the structuring of promotion meant that quite a lot of very good junior staff could not get on because there were no vacancies. In sheer frustration they decided to take their talents elsewhere and lots of people like Sam Anderson and Henry Wood left the ninth division to join companies like Wynns and Sunters which were better off for the experience and expertise which they brought with them. Very little could be done to encourage budding geniuses to stay for, when in government employ, you are paid for your position and not perhaps for your true ability.

All heavy haulage staff had seen big changes in the industry in every sense of the words. The heaviest electrical unit moved in 1952 was 150 tons whilst 16 years later this had doubled to 307 tons. Dimensions, too, were growing with petro-chemical distillation vessels being stretched so much that 120-150 ft was not unusual. Delivering them had prompted development of stronger tractors and trailers with the latest bogies able to be steered independently, if sometimes by brute strength. What had not changed a great deal were the roads which these loads trundled along. Alleged damage to roads led to hauliers being given a rough time by some sections of the Press. For those of us who can remember the Great North Road in its original form, which passed through every settlement between London and Edinburgh, one can sympathise with the car drivers who were incensed at having to crawl along behind so many slow-moving heavyweights. The development in the motorway age of the Preston bypass and the M1 did not do a lot for heavy haulage for sometimes height and weight limitations meant that big loads were excluded. With industry building more sophisticated equipment it was becoming less easy to construct things on site so, from all directions, pressure was being put on the heavy haulier both to carry bigger loads yet also to keep them off the roads. The 12 heat exchangers taken to Bradwell was a classic example of how to achieve both these aims but it was fortunate that the manufacturer and recipient were both close to the North Sea.

The trump card which the Ministry of Transport produced was the law. In enacting the Special Types General Order they gave operators virtual carte blanche to carry at will loads within certain dimensions, provided the requisite paperwork was completed correctly. Once these limitations were exceeded, viz a vehicle of more than 150 tons gvw, 90 ft rigid length or 14 ft 1 in in width, then a special order granted by the Ministry was needed. It was far cheaper to haul by road a 160-ton transformer from, say, Edinburgh to Southampton, but the Ministry order was more likely to dictate that the only bits of road to be used were between the factory and the docks in Edinburgh, then from the docks to delivery point in Southampton. Joining these two hauls up was an expensive ship, but it did mean that one excessively large load could be built and delivered with the least amount of road congestion.

'Vigilant', a crusader of the 'clear the road' campaigners, wrote in the *Newcastle Evening Chronicle* and his column regularly reported the noise, damage and congestion which heavyweights were causing in the north-east. What incensed him most was to see a rail locomotive being carried on a road-going low-loader, which was hypocrisy indeed. On April 25 1956 he heralded a new era as a 135-ton stator was delivered from C.A. Parsons, Newcastle to Littlebrook power station, Dartford, with the mileage between the Tyne and the Thames being covered on a coastal steamer.

Being the largest heavy haulier, the pressure was put on Pickfords to work hand in glove with the way the Ministry was trying to satisfy both the large load manufacturers and motoring public. Under that influence loads from Sheffield to Scotland, the latter always the most difficult part of the UK for heavy haulage, took rather a strange diversion. With about a dozen ultra-heavy castings required at the Ravenscraig steel works, Pickfords laid-down route was from Yorkshire to Liverpool and then by boat to the Clyde. Most of these were delivered without incident but one ship had the misfortune to collide with another vessel in thick fog as it was leaving the mouth of the Mersey and the castings sank to the sea bed never to be seen again. Even at Boroughbridge loads of steel were rarely lost without trace from the back of a low-loader, so on occasion ships do have their disadvantages. For those of you who are interested, the boat used on that particular occasion was the *MV Lurcher*.

County highways departments, always touchy about damage to their roads, started to complain over the use of solid tyres and apart from the heavy bogies, the 1960s saw the withdrawal of all the old Crane trailers. Even with pneumatics, bridges had to be regularly strengthened to withstand excessive weight so, if a haulier wanted this to be done, he had to ensure that the work was paid for. £5,000 went to one county council for just such a beefing-up for some heavy loads which were delivered incident-free into Scunthorpe. It was not until the job was finished that Pickfords found that only one side of the bridge had received attention and, as one might guess, all the loads crossed over on the weakened side.

Special order work for a haulier meant that he had to liaise closely with the Ministry, all the police forces and highways authorities whose areas he crossed, and also people like British Telecom, British Rail and, of course, the Electricity Board. Keeping all of these happy all of the time was sometimes not possible. Jock Fraser encountered such a situation as he drove Sunters' big Constructor into Carlisle one weekend in early 1963 en route from Workington to Consett with a 17 ft 3 ins wide ladle carriage, the all-up weight of which was close to 200 tons. The standard heavy load route through the city was too narrow so the Ministry-approved directions gave them a set diversion down Bochergate. 'You can't go down there' said the Local Authority, 'we've got road works.' 'This is a fine time to tell us' said Sunters, 'we'll just plate over the

holes and get past, it won't take long, we're used to doing this.' 'No you won't' said the LA, 'you'll wait till we are finished, we'll only be 3 or 4 weeks.' 'We can't wait three or four weeks' said Sunters, 'Let's be past.' 'You stay there,' said the LA. Sunters were furious for, as well as the load being required at CIC, the 150-ton girder trailer was booked for several jobs. 'Send me a load of timbers' said Jock Fraser and with no more to do the 80-ton lump was unloaded on to a city centre car park one night and the Sunter vehicles crept out of town. It was now the turn of the Local Authority to get upset and they insisted that the obstruction be moved but, until the route was finished and Sunters felt like recommencing the haul, the carriage went nowhere.

Pickfords had a similar confrontation in Manchester, not only with the council but also with the local police. They both agreed that the Regent Bridge just could not carry the 164-ton casting en route from Davey's at Sheffield to the docks for export to the USA. The Ministry pulled rank and said, 'That's our bridge, you go over it,' but the ensuing publicity meant that by the time the pull-push Diamond T combination reached Manchester, hundreds of people had gathered expecting something special to happen. The police had cleared everything and everybody off the bridge and all the attendants dismounted before the load started to creep over. Only the drivers were allowed to accompany the outfit and when it reached halfway the crowd held their breath. They were still holding their breath when it reached the other side for nothing had happened. At Ardrossan, however, something special did happen.

Getting large loads in and out of Scotland was always hard work. With height and weight limitations on the major routes, one of the standard heavy haulage diversions was the coastal route down through Ardrossan. The Scottish bridge authority had always hinted the Parkhouse bridge was weak but until about 1952, Pickfords never really knew what they meant by weak. 82 tons of bedplate and crankshaft were on the three-axled solid-tyred drawbar low-loader en route from Greenock to George Clarke's of Sunderland. Saturday night was spent parked up in Ardrossan and the following morning saw a little chain-drive box tractor head up an ex-WD Pioneer as the outfit climbed out of the town and turned right on the A78. It was still a climb up to the railway bridge so there was no rapid approach and the Scammells crossed at a steady pace. But when the two rear axles of the load carrier got on to the bridge, the roadway collapsed and trailer and load went down the hole. Even the tractors were pulled eight feet backwards before momentum was stopped as the trailer hit the railway line. Another state of devastation. It was the railways which came to the rescue on this occasion as two 80-ton cranes were quickly dispatched down the line from Motherwell and Rutherglen. In no time the load and trailer were extricated and apart from twisted drawbars the outfit was damage free. The hole in the road and damage to the crew's nerves took slightly longer to repair.

Strengthening the Scammell range in 1964 came the Contractor. Not a direct replacement to the Constructor, which stayed in limited production until it changed into an eight-wheeled freighter, but a more refined model with more options to the potential customer. With a choice of AEC, Rolls-Royce and Cummins engines, the 6 × 4 chassis suitable for either artic or drawbar use could also be fitted with either manual or semi-automatic transmission. Two basic cab options saw a crew cab very similar to the Super Constructor whilst the smaller version had smooth, rounded, pleasing lines. With the plated weight of the strongest model increased to 240 tons gtw, Pickfords soon brought them into service as the fleet flagships. Double, triple or quadruple use of tractors in combination

One of two loads of roof trusses moved by Sunters from Newcastle-upon-Tyne to Peterlee about 1967 gave no weight or width problems to the Mammoth Major, although its 125 ft length made it particularly impressive.

either pulling or pushing had been a way of life for a long time, however, unless the tractors were matched exactly and the drivers worked in close sympathy with each other, the transmission of each vehicle could take an awful hammering, it being impossible to ensure that changing gear and braking were done by all the drivers at precisely the same time. The new Contractor would not be able to carry everything single-handed but Pickfords felt that two of these pulling and pushing could tackle nearly anything on reasonable terrain.

In 1968 they pushed the record of the heaviest load ever carried to just over the 300-ton mark when they hauled a C.A. Parsons electrical unit from Newcastle to Dungeness. No direct haul by road was possible here, either, as the North Shields to Folkestone route was covered by a vehicle which was not suited to drive down the A1. The trailer used on this job was one of the biggest of the day, a 12-axle Crane rated to carry over 300 tons with an unladen weight of 89 tons. With a full load, mathematics will show that the axle loadings were well over the 30 tons a line mark, which made the Central Electricity Generating Board start to worry. This load might have been the first 300-plus-tonner but loads of this weight were to become fairly regular and some old bridges just could not take axle loadings of this kind. Ships were used to their maximum but everything had to be hauled by road eventually and, as there was a limit to the number of wheels which could be put under a girder trailer, something had to be devised quickly which could reduce loading to a more reasonable level. The answer was a load of hot air. Known as ACE 1, the air cushion equipment worked on the same principle as a hovercraft for, with the fitting of a flexible skirt and overhead ducting to the trailer, the activating of a blower truck attached at the rear forced air down on to the ground in the area between front and rear bogies. The trailer did not float away but the creation of weight transfer meant that, when required, the axle loadings could be reduced to a more tolerable 20 tons per line.

Twenty years after nationalisation the politicians did it again. True haulage, and especially heavy haulage, was near to being a closed shop with those who were 'in' ganging up and objecting to any potential competitors. But at a stroke the 1968 Transport Act threw out the A, B and C licences giving everybody an O licence which allowed all to carry anything for anybody, without restriction. In theory this meant that Joe Bloggs, who used to carry his own coal on a little 7-tonner, could sell his Ford and buy a 100-ton Scammell then go for traffic which had been strictly the domain of the select few. Prior to 1968 there had only been a few own-account operators in heavy haulage. People like George Wimpey and some crane operators required their own big low-loaders, as did British Nuclear Fuels Ltd whose cargo of little white boxes, similar to electrical units, was very heavy indeed. Starting with ERF artics the Cumbria fleet strengthened their potential with two Rolls-Royce powered Contractor-Crane Fruehauf combinations that were known by the locals as 'Coffin Carriers' because of the load they carried. Their standard haul was between Windscale and Chapelcross in Dumfries at an all-up weight of 96 tons. Not a lot of weight was actually carried but the radioactive qualities made it essential that sufficient heavy shielding was utilised. There was little likelihood of BNFL wanting to back load with the odd transformer but the change in licensing did mean they were legally entitled to do so.

1968 was a black year in the Wynns calendar for another reason. Running out of Stafford with a 130-ton load due to go into store at Hixon, circumstances found their police-escorted outfit on an unmanned railway crossing at the same time as an unstoppable train. Running into this sort of weight something tragic was bound to occur and, of course, it did. Whose fault it was is not a point to be discussed here although the resulting ramifications were food for thought for anyone connected with the industry.

The late '60s also saw the arrival of big continental tractors from people like Scania, Volvo and DAF which even in standard form were fit to run at 40 or 50 tons with little bother. Beef them up slightly and an operator had a tractor which was economical enough to run in general haulage yet, when required, had the strength to pull some fairly heavy traffic. Investing in suitable trailers was not a problem either, for this period also saw rental companies offer low-loaders for hire on a daily, weekly or monthly basis. This change in circumstances did not take away the requirement for the specialist heavy haulier, but it did mean that a lot of his bread and butter stuff was literally taken out of his mouth. The big two were obliged to react to this influx of competition for, although they could offer the experience of years, sometimes the customer was only interested in the price.

Right 415 DAJ was the MkII version of KVN 604, Sunters' first Mountaineer, which was close to being written off after a bad accident on the A6 at Shap in 1961. The outfit is pictured on Victoria Bridge at Thornaby with part of a rotary dryer being constructed at ICI Billingham.

Above and page 48 October 29 1963 saw this 21 ft 6 ins high cooler and flame trap en route from Head Wrightsons to the National Gas Turbine establishment at Farnborough. After squeezing past a Stockton Corporation Titan (see page 48), the vessel was shipped on the *Leven Fisher* from Middlesbrough to Southampton docks. Two days were required to finish the 30-mile haul, although on this stretch the leading Rotinoff fractured an air pipe leaving the Super Constructor to carry on unassisted. This reflected the sense in having two top capacity tractors employed on an important haul like this, the 90-ton load being well within the capacity of either unit.

Left Used to very different types of police escort, Sunter drivers were still taken aback when these two single horse-power mounts guided this vessel through Stockton town centre on May 29 1965.

Below The police escort waits for 447 DPY and one of the Juniors as they bisect the town of Wells in Somerset. Sunters' 90-ton capacity Crane 32-wheeler is shown in narrow form which makes it ideal for the carriage of crawler bodies, provided the tracks are chained up.

The Constructor was the backbone of the Pickfords fleet until well into the 1960s. At 70 tons this Ashmore vessel would prove little problem to the big 6x6 or the Crane solid tyred bogies.

Walsh's of Darwen seen making easy work of this crane section believed to be going from Dunkinfield to Consett in 1964. The well turned out Cummins-engined Atkinson configuration seems ideal for the job, although the two-trailer concept is rather questionable under modern legislation.

Barry's shop at Stockton got well used to looking at this 131 ft long stainless steel flash column which—on July 13 1965—was en route from Head Wrightsons to ICI, Wilton, because it took two hours to inch round the corner. Peter Sunter was quoted as saying, 'If the column had been just one foot longer it would never have got round, the operation was planned in great detail'. Not being one to exaggerate driver Robinson said 16 years later, 'It was a bit tight'.

Eight years after Bradwell, Sunter/Head Wrightson used the same concept of water-borne delivery for this 148 ft long column seen on site at Teesport. No big publicity for this but weighing in at 168 tons, it was at the time the heaviest load taken into the new Shell Oil terminal.

Total efficiency might be the heading of this photograph as the double headed Junior Constructors pause for breath and a fresh police escort on the trunk road near South Bank in July 1967. Sunters' solid tyred bogies are being used again and are shown to be ideal for the carriage of long heavy pressure vessels like this fractionating column en route to ICI Wilton.

Series One of the air cushion equipment seen 'blowing' on its first haul in April 1967 as Wynns deliver a 160-ton AEI transformer to Legacy, Mid-Wales.

Ten years after they were first introduced Siddle C. Cooks still felt the Junior Constructor was worthy of purchase, although by that time the power plus 680 engine was being fitted. The outfit is seen hauling a dockside crane cab section out of Clarke Chapmans, Gateshead, using two special 40-ton capacity Crane 16-wheeled bogies.

Late 1967 saw Sunters engaged in the delivery of this 120 ft long, 20 ft diameter, 95-ton fractionating column from Stockton to the Seal Sands chemical complex. Two Supers were needed to surmount Billingham Bank although, whilst the sedate pace did give the mate time to buy a pinta from the passing milkman, the drivers had to shout to be heard above the bellowing Rolls-Royce engines.

This well turned out Pickfords outfit is caught by photographer Alan Simpson negotiating Barry's corner at Stockton on April 26 1967. The big Constructor was more than capable of hauling this ladle en route to Consett Iron Company alone, but Pickfords were great utilisers of the little box tractor pusher in order to assist with negotiating tight turns.

STRATHCLYDE
171
Foden
KING
KING
FGM 93D

Left Looking like some overgrown mushrooms, these 23 ft wide 16-ton dome sections were hauled by Strathclyde Transport Services from Motherwell Bridge Engineering to Hunterston power station in February 1969.

A tight squeeze for the Rotinoff-Crane bogie combination on December 12 1968 with one of two identical Head Wrightson vessels which had to cross Newport Bridge at Middlesbrough. The original planning of the route suggested there were at least 3 ins of clearance but, in practice, there was a flake or two of rust dislodged.

Known as 'the Polish vessels' because of their destination, this load was one of five identical tanks 22 ft in diameter, weighing in at 95 tons, hauled to Middlesbrough docks in May 1968. This spectacular Dennis Wompra photograph shows Peter Clemmett inching the Rotinoff down Linthorpe Road, although not all the shoppers were greatly concerned over its presence.

You had to have a good head for heights if you worked as a mate in heavy haulage. Sunters hard worked Super is passing the Five Lamps at Thornaby on December 20 1968 with this vessel bound for south Teeside.

Above Wynns' first involvement in Nigeria was the origin of this outfit which was known as the Rolls-Royce road-rail wagon and is seen leaving Head Wrightsons in 1967. This company made the massive girders whilst the Crane Fruehauf running gear could be interchanged with rail bogies to suit that method of transportation.

Below Running at close to 25 ft high this Whessoe vessel created a fair number of problems in its journey from Darlington to Consett in 1971. The entourage of wire lifters and arm wavers are being kept busy as the Contractor-Crane combination turns off the A68 at Castleside not far from its destination.

5. Widening horizons

The expression 'the bigger they come, the harder they fall' was certainly true for Pickfords who, for over two decades, must have been the leading heavy haulier in the land. But come the 1970s the accumulation of circumstances saw them prompted to close 13 of their depots. It was perhaps slightly ironical that BRS had recently bought out the Tayforth group who in turn had owned Siddle C. Cook since 1964. The Consett fleet was allowed to operate as a separate entity for some time but, being quite close to Pickfords of Birtley, it was one of the first casualties and took the Cook name into history.

Having a lot more freedom than a nationalised company, Wynns' answer to the suppression of the '70s was to go even further afield in search of work. Entry into the Common Market might have been a reason for trekking into Europe, but it was more distant places like Kenya, Nigeria and the Sudan which were to be the hauling grounds for the Welsh men and machinery. Cyprus was their first port of call in 1971 when a new refinery was being built at Larnaca. The island does not have the general need for a heavy haulier, so when some large boilers had to be moved from Famagusta docks, Wynns' Pacific *Helpmate* and a trailer were laid on to meet the requirements. Nigeria always had close connections with Wynns since the mid-1960s when they had been approached by the Niger Dam Authority to devise an outfit which was to be hauled by road, rail or water. It was Head Wrightsons of Thornaby who made the girder trailer and rail bogies which were interchangeable with the Crane Fruehauf road-going axles, progress being maintained on land by Contractors front and rear.

Wynns' main project in Nigeria was the organisation of transportation of more than 16,000 tonnes of equipment from Lagos docks to the Ashaka cement works in Bauchi state, a haul of over 1,000 miles. The heavy moves were done by the Wynns fleet who found the distance and temperature quite a change from their small, cool homeland.

Similar working conditions were found in 1978 on the other side of Africa and it is probably in the Sudan that the greatest amount of pioneering was done by Wynns. One contract lasting over a year saw them haul 1,000 tonnes of 'out of gauge' refinery equipment the 1,400 miles from Port Sudan to Rabak, south-east of Khartoum. The convoy of load carriers was accompanied by a mobile workshop, living quarters, 'chuck wagon', radio-controlled general utility vehicles and a motorised scraper, the latter to smooth out the 'road' when required. It made a change from plugging up and down the A38 with cafes and telephones on every other corner, for when they crossed the Sudan there was a light aircraft in reserve, just in case.

The red brigade may have been the most famous of British trail blazers but others, too, were wandering off wherever required, Pickfords' Glasgow depot obviously being able to brag of the most exotic of heavy hauls. On the face of it, moving nine units totalling 600 tons would be fairly routine but when the location is up the Irrawaddy from Rangoon and it is the rainy season then things take on a different aspect. Glasgow supplied the men and a 100-ton MAN, whilst the customer laid on the trailer, a 20-ton winch and the problems. The men arrived with the temperature in excess of 100 degrees so they immediately felt at home as it was just like the furnace of a Scottish steel works. Plans of the intended moves were changed immediately and problem number one was how to get some transformers off a barge which was running in far too high for the slope of the basin. Simple, of course, remove the trailer rear wheels and use it as a ramp on which to winch the load off the barge, up the slope and on to drier land. Further up river a railway track had been laid into the bank side so that the equipment being unloaded there could simply be winched straight from the barge on to the rail bogies. However, with the river subsiding, the railway line was too short, the mud was too deep, the rain was too heavy and, to cap it all, the customer's winch broke. Ye of little faith can still rely on Pickfords, for the enterprising crew found an ancient Scammell boom wagon just lying about waiting for such an emergency. With the fitting of a new wire winch rope the problems were surmounted and the loads delivered, of course. The men and MAN returned to Scotland although memories of this haul would never be forgotten.

It is Pickfords Rugby branch who must brag of having the fastest moving heavy haulage trailers in the land. The Crane Fruehauf drawbar trailers look fairly innocuous and on the memorable move they were carrying just over 30 tonnes each of electrical cylinder bodies. Bound for Taiwan, their customer decided he wanted them as soon as possible, so flying was the only answer. A USAF military transport C5A Galaxy, nicknamed *Fat Albert*, an example of the largest cargo aircraft in the world, swallowed up the trailers and loads after being expertly pushed up the 1 in 4 loading ramp by a ballasted Atkinson. Leaving Mildenhall on the Sunday, it flew via Alaska and Japan arriving in Taiwan three days later. It is amazing what tricks heavy hauliers get up to.

For those with their feet on the ground, the early '70s meant North Sea oil and Nicholas. The modular trailer was here to stay and although the German company Scheuerle were the early leaders, it was the French Nicholas concern who were to come up on the rails. Cranes of Dereham had produced some small-wheeled bogies but obviously with a limited national market it must have been difficult to justify investing heavily in the research of this new concept. Kings produced some six row bogies for Pickfords but unfortunately these were not without their problems, thus it was open to the Frenchman who seemed to get everything right. Two basic versions, light or heavyweight, were offered and, as well as being able to add on axles ad infinitum lengthways, some trailers were made to be split in half widthways. With suspension and steering of all wheels down to highly sophisticated hydraulics, these new load carriers were the answer to nearly every heavy hauliers question. They were expensive but it did mean that if a company bought, say, 20 rows they could have any permutation of these axles lengthways or sideways which made them adaptable for nearly every conceivable job. Nicholas also produced trailer beds and adjustable goosenecks so general heavy haulage arti s could also be purpose-built in true Meccano fashion.

Modular trailers were soon in demand as the heavy engineering companies began building hard in the race for oil. Assembling a rig deep in the North Sea needed entirely different building techniques to suit the environment. Once the framework had been sunk in situ, it was ready to receive its fittings which had been preassembled on shore like giant Lego building bricks. It was a straightforward matter for the deep sea crane to lift these off the barge and place them in position but to get these modules on to the barge the new trailers were necessary. It was not practical for every construction site to have suitable craneage to lift anything up to 2,000 tonnes, so the demand was made for the specialised heavy transporter to enter this field moving items around the site or on to barges and ships which, in turn, ferried them out to the rigs. Both Pickfords and Sunter Brothers were ideally placed to service the northern yards but, as the heaviest loads they had moved in the past were around the 300-ton mark, the prospect of moving items nearly ten times that weight was rather frightening and initially they shied away from this work as they felt they were out of their depth.

Being such a specialised entity, 'load-outs' prompted the forming of an entirely new company in the heavy hauling scene, the aptly named Magnaload. Owned jointly between Peckstons of Middlesbrough, who were in the shipping business, and Mammoet Transport BV of Amsterdam, probably the biggest heavy haulier in the world, its three-vehicle fleet of two 6 × 4 Volvos and a second-hand Contractor would hardly overawe the big two, but virtually the first haul it made put it into the record books. Four modules weighing in total nearly 7,000 tonnes were loaded-out on to barges in the River Tyne at the William Press Howden yard at Wallsend and the tractors did not even break into a sweat.

Archimedes would be the first to agree that you cannot just push a 2,000-ton lump on to a 300 ft long barge without something happening, so, to compensate for his principle, numerous pumps moved around the pre-loaded water ballast into different compartments inside the barge and eventually back into the river. As well as the transfer of weight, due compensation also had to be made for the rise in tide, it being unthinkable to try and load-out with the tide falling. Magnaload's method of moving this sort of weight slowly on to the barge or around the site relied on the strength of winches either mounted on the back of the tractors or welded to the floor of the barge with at least two pulling whilst another two were releasing, but always being ready to brake or retrieve if the circumstances dictated. Magnaload were able to offer the know-how for managing moves like this and being able to call on the equipment owned by the Mammoet organisation, which was anything down to tugs and barges, a lot of work was to come their way.

Rigging International, an American-based company, were also active in this field but their approach was slightly different in that they preferred to use crawlers. Looking something like the base of a large crane, these self-contained units were simply driven under the load and carried it about to wherever it was wanted, thus cutting out the need for trailers, tractors and winches. One of their drawbacks, however, was that they did not have hydraulic suspension, so loading and unloading were laborious separate jacking procedures. It was to be this handicap which persuaded Rigging to sub-contract some of their moves, so allowing Sunters a taste of heavy haulage, load-out style. Transporting this sort of weight was not difficult to do, if you knew what you were doing. The Dutch and Americans knew how to do it, as did Messrs Duffield, Wilson and Pearson.

These three ex-Rigging employees pooled their money and set up business offering for sale their expertise in the load-out field under the banner of ITM. They did not have any equipment at this time so, in the first instance, they went to Pickfords to hire the men and modular running gear to do the actual physical work when the loads were to be moved. But it was the structure of the blue brigade which was to create difficulties with ITM. Having so many depots, their new Nicholas axles were spread country-wide, so getting 30 rows together all at the same time for a load-out meant bringing eight rows from Birtley, ten from Glasgow, six from somewhere else, and so on. Once they had been assembled on site they had to be rigged up and tested which in itself was a lengthy process so, to a go-ahead young company like ITM, Pickfords seemed just a little bit slow. Unlike Pickfords, all Sunters' new axles were based at Northallerton so producing 20 or 30 rows was little problem and, of course, if desired Wynns could help out if they ran short. Peter Sunter had been champing at the bit trying to get back into the load-out field when Sunters and Rigging had both agreed to part company so, when the chance came up in 1977 to inaugurate the Sunter-ITM combine, he leapt at it, just in time to bid for the Shetlands.

The giant oil terminal being built at Sullom Voe presented a challenge to all concerned, BP the terminal manager acting on behalf of 32 participant oil companies and Constructor John Brown, the managing contractor. The extreme weather conditions, combined with the sheer isolation of the huge site, virtually dictated to CJB that wherever possible the major units for installation should be prefabricated. In fact 40 per cent of the process facilities were pre-assembled in module form weighing anything up to 500 tonnes each and it was later calculated that two million man hours had been saved on site by utilising this form of building concept. Sunter-ITM's first contract was for the delivery of 37 piperacks which were designed as linkable modules each approximately 72 ft long, 50 ft wide, 65 ft high and tipping the scales at 300 tonnes. The delivery timetable was highly concentrated and involved up to six individual racks being loaded-out on to one barge at the same time. From Leith, Tyneside and Teeside the flat-topped carrier set out northwards but such was the delicacy of the racks' specification, they could not tolerate any longitudinal deflection (they must not be allowed to warp) greater than 18 mm which, to you and me, means less than an inch. To compensate for such problems this form of heavy transportation can be very lucrative indeed. The distances involved are very short, but the weights carried are astronomical.

Following their success in the Shetlands Sunter-ITM went from strength to strength in the load-out field. It was mainly North Sea oil work, but one slightly different job was the moving of the Thames barrier gates from the Cleveland Bridge fabrication yard on Teeside. Not particularly high, the four main gates were nearly 200 ft long and weighed in at a hefty 1,500 tonnes each but the hydraulic suspension of the Nicholas trailers proved ideal to lift, carry and unload them on to the moored barge with little difficulty.

Since 1964 Sunters had always seemed to be the junior partner when compared to Wynns in BUT heavy haulage, but entering the '80s when the recession in road haulage began to bite, it was the Northallerton company which was to be the group's shining light. 1980 saw the SB-ITM combine take their expertise to the Persian Gulf. Sufficient contracts were won to justify a respray and a long journey for Contractor HVN 397N and a 6 × 4 Volvo tractor plus 20 rows of Nicholas running gear. They looked more like vehicles of a UN peace-keeping force. Gone was the distinctive grey and maroon colouring; white was now the predominant colour. A round badge on the door of the vehicles proclaimed they were part of the Hercules International Transport Co Ltd, this being a joint venture set up by Sunters, ITM and an agent from Dubai. One of the jobs the combine did warranted high precision know-how and was just a bit beyond the usual Arab or even British haulier. This entailed the moving of six massive balls or, to give them their correct title, pressure surge spheres, from the CMP fabrication yard at Ajman via barge to Zirku Island and then a 2-mile haul to their destination. Each was 200 tonnes in weight and 53 ft in diameter so that even the mighty Contractor was dwarfed by its charge. The trailers were fitted with a cocquetier to accept the radius of curvature of the sphere, it being specially arranged so that pressure was equally distributed and damage to the sphere's fittings avoided. Loading and unloading was done solely by using the stroke integral in the Nicholas suspension and the final deposit was done to such fine tolerance that it discounted completely the need for any heavy lift craneage. Yet another notch on the SB-ITM handle.

As Sunters and ITM were going from strength to strength, the fate of Magnaload took an interesting turn. Their destiny should have been sealed when Peckstons, one of their joint owners, went into liquidation but one Tom Llewellyn decided he might be able to do something about it. Since he had taken over as managing director of Econofreight Transport in the early '70s he had changed both the image and traffic load of this Transport Development Group company. The general haulage vehicles were still there but the tippers had gone and in their place were long load and heavy haulage outfits. Also operating out of Teeside he was well aware of the prowess of Magnaload. It was certainly a very small fleet, but it was the quality of their equipment and the know-how of the staff which held great appeal, so out of the ashes of Magnaload in 1980 came the joint venture of Mammoet-Econofreight. Initially a 50-50 set up, but as the Middlesbrough staff found their feet, two years later the name had not changed, although Econofreight was the predominant partner. Their operation was similar to that of Sunter-ITM in that both companies helped each other either with equipment or expertise but Mammoet could also offer hauling facilities for traffic destined for Europe and their sea fleet was best described as large.

Teeside was thus the base of two of the heaviest movers in Europe. From humble beginnings ITM had developed into a company which Britain could be proud of, they may have respected Mammoet but were not overawed by them. The change in circumstances have meant that the big two of the '60s have now been replaced by the two massive movers of the '80s. Pickfords and Wynns may have been overshadowed but they were not written off.

Below left Hauling 511 tons in three lots from Birkenhead in 1971 was a big job even for Wynns for, at the time, they were the largest loads ever moved by road. Up to 28 ft in diameter, 111 ft long and 212 tons in weight, they required most of the road and a lot of police attention in the 17-mile haul to the Shell refinery at Stanlow.

Below right Seen approaching Newport Bridge, these 28 ft wide 110-ton half sections of a converter were quite a handful. Destined for the BSC Anchor project at Scunthorpe, the motoring public were bound to be pleased that between Middlesbrough and Lincolnshire they were carried on board the double hulled *Gloria Siderum* thus cutting traffic chaos to a minimum.

Above left South Wales is a testing place for heavy haulage. This 220 tonnes stator from AEI Manchester needed four of Wynns' biggest tractors to get it up the 1 in 6 Buttrells Hill when being hauled from Barry dock to Aberthaw power station.

Above right British Nuclear Fuels operated two of these Rolls-Royce-powered Contractor-Crane Fruehauf combinations in the '70s out of Windscale and Chapelcross. Because of the hazardous white boxes they carried, West Cumbrians nicknamed them 'Coffin Carriers'.

Below During the mid-1960s, *Dreadnought* received a massive face lift and had more than its share of Scammell bits under the Pacific skin. It is seen hauling seven rows of Nicholas bogies supporting an old loco destined for preservation.

One of Wynns 300-ton capacity Crane trailers seen entering the Davy Roll Company at Sheffield in August 1974 with this 195-ton casting from Germany. The three tractors, headed up by the modified *Dreadnought*, had to work hard crossing the Pennines from Manchester docks with an all up weight close to 400 tons.

One of the record-breakers carried by Sunters was this 228-ton steel column having an impressive length of 174 ft. November 10 1975 saw HVN 397N and 14 rows of Nicholas on its fairly short road haul from Tees dock to ICI Wilton.

Below A good example of the modern day Pickfords heavy haulage outfit — Contractors pulling and pushing TM 1120, 12 axles of 300-ton capacity Crane Fruehauf trailer. Seen in 1973, not exceeding the speed limit, they are taking one of 12 167-ton low pressure steam turbine rotors from GEC Manchester to GEC Rugby for overspeed testing prior to export.

An early experience for Sunters in the load-out field was the site haul of eight modules each approximately 75 ft × 35 ft × 26 ft and 380 tons in weight. The odd layout of the Nicholas running gear was needed to turn the vessels prior to them reaching the water edge.

An early Detroit-powered Contractor artic of Heanor Haulage shows how the modular semi-trailer can carry excessively heavy loads, provided the weight can be placed over the axles.

Made of Wolsingham Steel, shaped by Sunderland Forge and hauled by Olivers Fodens, this 40-ton ship's rudder is seen leaving the Wearside works en route to Brazil.

Above left and right Sunters Titan I and Nicholas tri-axle semi-trailer is seen in 1978 with two very impressive loads. VVN 910S was made by Titan Gmbh of Appenweier, West Germany and the outfit cost Sunters in excess of £100,000 when bought new in 1977.

Centre left and left XUP 999F was the last big Scammell to be bought by Siddle C. Cooks of Consett before they disappeared from the heavy haulage scene completely. It ended up with Northern Ireland Carriers, a Pickfords offshoot, and is seen in December 1977 helping Magnaload into the record books. Built by GEC Larne, this 120 ft long 401 tonnes moisture separator-reheater vessel contained 35 miles of tubing and, when bound for San Onofre nuclear power station in California, it was the heaviest load moved on roads in the United Kingdom.

Above With the choice of either paying £15,000 for the removal of street furniture in Newcastle or £1,100 for the exclusive use of the Tyne Tunnel, Econofreight obviously chose the latter method in order to get this 50-ton Richardson of Darlington fabrication north of the River Tyne. This photograph, taken at 2 o'clock in the morning, shows what a tight fit the load was. Econofreight knew they could get it through, however, because with good foresight and planning they had taken a wooden replica modelled to the exact dimensions of the real load through without any trouble, although driver Steve Ford only had 3 ins clearance on either side. Doing the haul is the much modified DAF 2800 which is fitted with an Allison automatic gearbox, hauling Nicholas axles.

Below Middlesbrough transporter bridge, itself a great invention, looks down on this Econofreight outfit seen about to leave for North Wales. Getting these empty boxes to Air Products at Wrexham was impossible until this concept of end suspension was devised; although there was no weight problem for the ballasted 110-ton capacity DAF.

SIF
HOLLAND
HOLLAND
MAGNALOAD
TRL
924H

Above At 370 tonnes, this vacuum distillation column was the heaviest and most impressive of the loads hauled into the Lindsey oil refinery by Pickfords in August 1978. Its 40 ft height necessitated the removal of 42 lamposts and the raising of an NCB conveyor belt 21 ft to afford passage.

Left These spectacular Golf Oil photographs show a very awkward 220-ton vacuum distillation column entering the Milford Haven refinery in 1977. Eight double rows of Scheuerle running gear plus two thirds of the Magnaload fleet were needed for the job due to the all up weight being in excess of 300 tons.

Below Roy Brandley keeps a watchful eye on this awkward vessel as he steers the Nicholas on to the wrong side of the dual carriageway at Stockton, late in 1978. Econofreight like the four-deck-four trailer and use it a lot when rigidity between bogies is required.

Above Pickfords are probably the most famous name in house removals but Econofreight show how they can do it in a slightly different manner. The automatic steering on the Nicholas semi-trailer is shown to good effect as the outfit rounds Stranton roundabout in Hartlepool late in 1978.

Above left and left 712E was an example of the first batch of 240 Contractors to be operated by Pickfords after their introduction. The Scammell is easing out of Foster Wheelers at Hartlepool early in 1979 with a waste heat boiler destined for Immingham, all up weight being close to 250 tons. Left-hooker 925T, a MkII version, was only one week old when it headed up to ease passage over Newburn Bridge.

Above The MACK Interstater found favour with quite a few operators in the middle range of the business, Baldwins of Stockton operating two of these 100-tonners. They mainly used them in the carriage of their own crane equipment, but the openess of the O licence meant they could haul for others when site operations were quiet.

Right TSL get good service from their fleet of Volvos, this F89 making easy work of the NCK crawler crane en route from Middlesbrough to Essex. The King semi-trailer did need a helper tag axle to spread the weight of 89 tons which the combination was grossing.

Right With all the big Sunter tractors employed elsewhere, Wynns were asked to help out with a short haul to Hartlepool docks of this 180-ton Foster Wheeler package boiler in early 1979. Bound for the Texaco refinery in Pembroke, the load required very careful carriage which meant an easy time for *Invincible* and a cold crawl for the police escort.

Left March 1979 saw Sunters aptly named *Fearnought* engaged in the haul of one of three silos to ICI Wilton. Although the vessels had been manufactured at Dock Point, Middlesbrough, they had to be transported down the Tees by barge to Teesport so that the eventual road haul was kept to a minimum. The Cummins-engined tractor was hauling 16 rows of Nicholas bogies with 256 tyres to support the 98 ft long, 50 ft diameter vessel, all up weight being 343 tons.

Below Wynns 13-year-old Contractor heads up Whessoe's younger Scammell on one of the testing gradients in the heart of the Elidir mountain hauling steel lining pipes, an integral part of CEGB's Dinorwic power station, North Wales, April 1979.

6. Horses for courses

Into the '70s the Contractors became established as the leader in the UK heavy haulage locomotive field. Even Wynns returned to conformity in 1966 for, when their hard-worked Pacifics and Diamond Ts required replacing, it was the new Scammell range which took their place. *Dreadnought* received a massive face-lift and a new registration about this time but, although it continued to work hard for its keep, its days were numbered. The manufacturer's plate affixed to the strongest Scammell said that it was designed to haul a gross combination weight of 240 tons, but what it could actually pull safely was anyone's guess. The makers were not interested in changing the figures on the plate, although they liaised closely with operators who wished to operate at weights in excess of this amount. This seems a strange statement to make but two vessels moved by Sunters in February and April 1977 were a point in question.

Manufactured by Foster Wheelers of Seaton Carew, their eventual destination was only a few miles south on the big Seal Sands chemical complex but, owing to their massive 376 tonnes weight, they had to be road-hauled north to Hartlepool docks then ro-ro'd back across the Tees bay. If the tonnage was not bad enough, each vessel measured 92 ft long and 31 ft in diameter, dimensions which meant drastic modifications to the street furniture on the 4-mile road haul. One of these alterations was the swivelling round of all the street lights so that instead of shining on to the roadway, they faced outwards and illuminated the grass verges. With only two months separating the hauls of the two, the streetlights were just left in this position after the first job was done. Motorists did not suffer a great deal from any lack of light, but they did make numerous 'phone calls to the local police to report that 'vandals' had turned round all the lamposts on the Seaton Carew road. To get the best stability and weight distribution, Sunters' engineers decided to utilise two Nicholas bogies but, following load-out techniques, the 2 × 7 row configuration ran at a width of 19 ft and had 12 tyres per row. They were, in fact, an example of the Nicholas trailer and a half running side by side whilst a 1 ft spacer held things in position. Adding everything together the all-up weight including tractors was 634 tons, this was, of course, a calculated weight for even Cleveland Police would have been hard pushed to find a suitable weighbridge to check it. What pulled them? Contractor LAJ 798P with Bill Jamieson at the wheel, all on its own. True, Peter Clemmett in NAJ 103P did double-head over Newburn Bridge, but this was just as much to control the descent as to aid in surmounting the initial incline. Watching the Contractor travel along the Tees Road, the serene ease of its passage conveyed the impression that it was pulling something like 60 tonnes instead of over 600. In fact, there was far more smoke coming out of the cab window from burning tobacco than was spotted leaving the Cummins exhaust pipe. Newburn Bridge and the docks apart, the hardest point in the haul must have been the actual leaving of the factory. This, and all other 90-degree turns, saw the big driving wheels dig in and push hard, trying to straighten out the front wheels which had to scrabble hard to hold directional stability. Mainsforth Terrace in Hartlepool was closed to traffic when the load took up residence. At nearly 40 tonnes a row or over 3 tonnes a tyre, the negotiation of the Church Street railway crossing was very delicate and had to be timed for the early hours when there was a lack of opposing traffic.

The Contractor was a good tool, make no mistake about it. The steering lock could have been better and when running empty it was a bit leisurely, but for pulling the weights it was hard to beat for strength. Scammell knew they had a winner and there were numerous examples about to show that they were not a nine-day wonder and would last as long as you wanted. December 1977 saw Magnaload haul what was then the heaviest load ever pulled on

UK roads. At 401 tonnes, the 121 ft long moisture separator-reheater vessel was one of four similar loads made by GEC Larne bound for the San Onofre nuclear power station in California. For the haul to Belfast docks, the 16 rows of Scheuerle bogies were hauled by TRL 924H, the second-hand Magnaload Scammell, ably assisted by XUP 999F, which had been hired for the trip. This latter Contractor had started life with Siddle Cooks of Consett but, come the takeover, had eventually ended up with Northern Ireland Carriers, an offshoot to the Pickford empire. The two old tractors thought nothing of the 25-mile journey even though the all-up weight was 635 tonnes. Pickfords expect, and get, a long life from their fleet of Scammells and show little respect for their age. When they moved 730 tonnes in three lots from Immingham docks to Lindsey oil refinery at South Killingholme during the summer of 1978, the most impressive load must have been a 370-ton column which was over 40 ft high and 120 ft long. A National Coal Board conveyor had to be raised 21 ft to allow it to pass and there were 42 lamposts removed en route, by workmen not by the load. Pulling this vessel and the Nicholas running gear was SYO 384F, certainly ageing but not aged.

Prior to the Mammoet arrangement, the biggest tractor in the Econofreight fleet was UVN 44S. This well turned out Scammell worked hard for its living but very few people knew that it had already spent one lifetime working for ICI in Cheshire. Sunters also rebuilt one of their first Contractors, TPY 675H, after eight years of work. At a cost of £42,000 it shows the job was not skimped in any way, but it also shows what an operator will do rather than pay what he feels is Scammell's rather high price for a new one. The manufacturer may not have to worry a great deal when an operator is wanting a tractor to move over 200 tonnes for he is close to running a monopoly, but when the weight comes down a bit quite a few people, rather than go to Watford for their machines, have pulled out some interesting vehicles, most interesting of which must be the Searson alternative.

As one of the new generation of heavy hauliers, Peter Searson, MD of Heanor Haulage, just would not accept it when he could not get his ideal choice of tractor either because the manufacturer could not or would not build what he wanted. He did not particularly like the big Cummins engine nor did he want the semi-automatic transmission but when the conformists said there was no alternative he decided to show them there was. His reply was ONN 686P, an eight-wheeled tractive unit which looked a bit like a Volvo, a bit like a Scammell but was really a Heanor Haulage built HHT. The heart of the vehicle was a Detroit 8V92 engine pushing out 400 bhp and, to anyone who appreciated the sound of the old Foden two-stroke, it made a noise like Mozart. A special engine powered through a special 15-speed Fuller gearbox, via a Spicer clutch to an interesting rear bogie. Three Leyland axles were used to take a heavy loading from the semi-trailer but, in order to assist with manoeuvring, the foremost axle was not driven and could be lifted hobo style. To top it off a Volvo F88 sleeper-type cab was utilised which meant that with the engine compartment being outside up-ahead, there was ample room inside.

HHT 2 soon followed, although this one was of conventional 6 × 4 configuration whilst a splitter box was fitted offering over half a century of gear ratios. With plans to have five of these machines in his fleet, Peter Searson has no intention of putting Scammell out of business but the HHT's have already acquired a reputation of being able to pull the weight as well as being able to traverse the ground at a rapid pace.

Peter Sunter also came up with an alternative when he was searching for something to fill a gap in the North Yorkshire fleet. Scania and Volvo were already well represented and four Contractors formed Sunters' backbone but the outfit which took to the road in June 1978 was something entirely different. VVN 910S was its number and nearly everyone referred to it as a Mercedes as it passed rapidly by. Well it did have a Merc cab, a three-pointed star on the front plus a lot of Benz bits, but its true maker was Titan GmbH of Appenweier, West Germany. The making of the Titan was prompted by one Heinrich Schutz who was probably the German equivalent of Peter Searson. He, too, was frustrated when he could not get the heavy haulage tractors he wanted, so he approached Titan Special Purpose Vehicles Ltd who made things like fire engines and forestry vehicles. The result of this get-together was the Titan tractor but, unlike the HHT, there have been over a hundred of these produced and sold. Sunters' Titan I was a 6 × 4 version powered by the 21-litre Mercedes OM402 V12 engine, unturbocharged but still producing 420 bhp. Torque was 1,036 ft/lb at 1,600 rpm but the big thing it had was a ZF transmatic transmission. The eight-speed plus crawler synchromesh gearbox was driven through a WSK 400 torque converter so it meant all the power on tap could be put through the wheels and traction was assured. Lack of traction had been one of the consequences of the fluid flywheel of the SCG semi-automatic gearbox fitted to the Scammells. The Contractors suffered at times as not enough power could be transmitted when working at low revs, whereas the Titan would just walk away from a situation on tick-over with the torque converter multiplying the power up as required. The Titan was rated to run at up to 200 tonnes as an artic or 400 tonnes gross as a ballasted tractor whilst Ken Bickerton, its normal driver, regularly coaxed 80 on to the speedometer (km/h, of course) which reflects on the spread of performance which the vehicle could offer.

Titan II, EJW 229V, joined the Sunter fleet in February 1980. Nearly identical to its predecessor it did, however, sport a 6 × 6 configuration and was bought primarily as a ballasted tractor. Titan II's only 'fault' is being unable to get enough weight on to the front axle for, once this starts to spin, the outfit has to be stopped and all the diffs locked up before progress can be resumed. Cats-eyes, of all things, have a habit of knocking the axle into spin, so they are treated with great respect; not that restarting the vehicle is much of a worry as Malcolm Johnson found on the steep climb away from Penrith on the A66 near Stainton Village. More concerned with watching passing talent, he did not realise the Titan required a few more gears in hand and came to a gentle halt. There was 120 tonnes of concrete beam sitting on the back en route for Windscale but that made little difference and the outfit just eased away with the torque converter in full control.

Sunters' only question mark over the Titans is whether they will stay the pace as well as the Contractor but, of course, only time will tell. If they ever doubted that the vehicles could pull this was demonstrated at William Press's yard on Tyneside when Ken and Malcolm hitched the Titans up to 98 axles fitted with nearly 800 tyres supporting a production module destined for the Fulmar field. All-up weight was close to 2,400 tonnes and the two tractors moved it across site prior to being loaded-out. Yes, 2,400! The oil temperature gauges of the converters started to rise a bit but they were seen to be man enough for the job. Put the MkI Contractor on the open road where the throttle could be opened wide and it would probably out-pull the Titan, but on tick-over inching about a site, the torque converter nearly reigns supreme.

The Titans may have been different but they were at least recognisable which was something which could not be said about the next big locomotive which came to Northallerton in 1981. If you had to guess you might say it was a Ford or Berliet and, in fairness, the cab was from Renault. The engine was the massive 450 Cummins, the whole machine being the product of Nicholas of France, the trailer maker, and went by the name of Tractormas. The beast is yet to prove itself but, with a Clark torque converter and gearbox in the transmission line, its potential is very promising indeed.

The greatest user of home-produced tractors must have been Pickfords who, since the '50s, have shown a predominance of Scammell in their fleet. The National Freight Corporation were probably instrumental in the production of the Crusader for general haulage purposes and Pickfords, their heavy division, got good work from the stronger versions. 1970 saw an eight-wheeled artic tractor unit produced with the Motor Panels cab, Detroit engine and the name of Samson but only a handful of these were made. For 65 tonnes gross, the standard 6 × 4 Crusader was ideal but for 100-150 tonnes, Pickfords felt that use of the Contractor was an under-utilisation so, in search for something to fit, they went to MAN for their Jumbos. Versions of the 32.400 and 38.320 went into general service but when a 40.400 came over as a demonstrator in 1979, this was felt to be an even better machine and CYL 510V, an example fit for 180 tonnes based at Glasgow, is proof of the pudding.

Most favoured of the imports are undoubtedly the Swedish makes of Volvo and Scania, who have a great following in the lighter end of the business, not least with Mike Cave. Nine years after setting up shop in 1973 he has developed his one-man band into one of Europe's most respected heavy haulage companies. Based at Bedford, Planthaul Ltd specialises predominantly in overseas work travelling throughout the Common Market and as far afield as Egypt and Libya. Their immaculately turned out fleet is at present all Volvo, the flagships being three very special F12 6 × 4s, whilst the remainder consist of 13 assorted F12s and four F7 4 × 2s. With all the loads being driver-accompanied right through to their destination, the spares and service network maintained by Volvo in Europe is obviously a great attraction to a company in Planthaul's position. The trailers are also predominantly imports with Goldhofer and Broshius held in great esteem.

Planthaul operates solely artics but the SKPH ten-axle Goldhofer with hydraulic suspension, automatic steering and load compensation means that up to 140 tonnes can be carried without overloading the king-pin or the tractors' drive axles. The latter is a problem which faces all heavy hauliers, the modern trend being wherever possible away from tractors and drawbar trailers which carry a heavy penalty of ballast. The sophisticated hydraulically suspended modular semi-trailer where axles can be added to order is ideal if the weight can be placed over these wheels, but for longer, awkward loads it may be impossible to keep the king-pin loading down to an acceptable level.

Low-loader operators in some other countries utilise two swan necks on their heavy artics with the first step frame semi-trailer being very short and only intended as a means of getting two or three more axles under the load. But, with UK legislation preventing the adoption of such tactics, Robert Wynns was one company which was trying to come up with some bright idea that would produce the same result. They also wanted an outfit which could carry 100-110-tonne loads but which would not exceed the magic figure of 152 tonnes gross taking them into special order work meaning greater Ministry control and expensive ship utilisation.

Sam Anderson went to Nicholas of France in 1978 with this problem and he returned with a design for a 150-tonne artic. The semi-trailer had two sets of axles mounted immediately behind the adjustable swan neck and before the commencement of the bed, whilst the remaining four rows followed at the end of the trailer. Automatic steering of all axles saw the front two rows steer in sympathy with the front axle of the drawing tractor whilst the back four turned in the opposite direction. This meant the turning circle of the outfit was relatively small as it seemed to handle more like a small rigid than a big artic, something the drivers had to remember at all times. The new concept answered the original questions although it did have its fair share of teething troubles, especially over the steering of the first two axles and adjustment of the weight distribution through the swan neck. Pickfords' Glasgow depot put one to work behind one of their big Jumbos whilst Dawsons of Middlesbrough used a bonneted Scania to haul theirs. Brackmills bought one, as did Heanor Haulage, but the latter were soon to sell theirs to G.C.S. Johnson, an up and coming heavy haulier operating an all-Scania fleet from Barton near Scotch Corner. Wynns had the largest tractor of the lot hauling their new semi in *Renown*, a large cabbed MkI Contractor modified slightly with suitable tyre equipment, based at Manchester.

The Wynns heavy haulage fleet of the early '80s had seen a big change from that of 25 years earlier with the imported heavy tractor now being the exception rather than the rule as the fleet was predominantly Scammell. Bedfords were used for the lighter loads whilst Wynns had a version of the 6 × 4 Crusader called *Amazon* which was fit for 100 tonnes gross operation. Both light and heavy versions of the Contractor were utilised and Wynns were probably the only UK operator to use these big Scammells generally in articulated form. In fact in the Sudan, powerful *Hercules* regularly ran as a double bottom, an artic pulling a drawbar trailer. Wynns had not got into the heavy load-out field as much as Sunters, although utilising these techniques they had regularly moved anything up to 500 tonnes from or on to adjacent shipping. They did not have the great demand for torque converters on their tractors but, when Scammell offered this option on the MkII Contractors with the bigger 450 Cummins engine, *Invincible* and *Superior* came to enhance the fleet with this specification.

Wynns were reckoned to have been involved in practically every heavy engineering exercise in Wales since the late 19th century, the most unusual of which must have been the construction of the CEGB Dinorwic pumped storage station inside the Elidir mountain, Llanberis, North Wales. The scheme puts into use two lakes, one at the top of the mountain and one towards the bottom. When power is needed, the tap is turned on and the top lake releases water which is channelled through tubing to the bottom lake and generates electricity by driving turbines in a similar manner to that by which numerous hydro-electric schemes operate. To ensure Dinorwic never runs out of water, during the night on cheap rate electricity the contents of the bottom lake are pumped back up to the top ready to supply instant power once more to the national grid as and when the requirement is made. What will probably most please environmentalists is that when the scheme is completed there will be very little to damage the beauty of Snowdonia, as everything is concealed inside the mountain.

Chris Millers of Preston were involved in moving lots of the turbine equipment from the manufacturers, Markhams of Chesterfield, to site but, such was the geography of North Wales, that most of the traffic had to be hauled between Manchester and Port Penryn by ship. Amongst the heaviest loads were six valves used to control the flow of water out of the top lake and, at 150 tonnes each, they must have been some of the heaviest taps ever made. To suit these big taps, Millers operated what was believed to be one of the biggest Macks in the world, a 6 × 4 ballasted tractor rated for 240 tonnes gross operation. The Bulldog made the greatest impact of all the American trucks in the UK market of the late '70s with quite a number finding their way into heavy haulage operations. The maxidyne engine was renowned for its pulling ability and general haulage versions rarely had anything more than a five-speed gearbox such was the spread of performance that the flexible engine could offer but, to ensure it lived up to its rating, Millers' Mack had 12 gears in its transmission line. The loads were carried by Cometto modular running gear, ten rows and a set of girders utilised to carry the 14 ft wide valves. Millers made great use of the hydraulic suspension on the Italian trailers for picking up and depositing their charges, an asset also greatly appreciated by one heavy engineering company turned haulier.

Whessoes of Darlington were contracted to produce 4,600 tonnes of steel pipes ranging from 8 ft 6 ins to 14 ft 9 ins in diameter and in sections weighing up to 130 tonnes, for installation inside the mountain. Transportation from County Durham to North Wales was put in the capable hands of Sunters and the Fisher boats but actually to place the linings in position, Whessoes bought one special outfit and did it themselves. The locomotive part of the combination was XDC 485S, a Contractor, but with 32 tonnes of ballast on its back it tipped the scales at 50 tonnes gross. The special trailers were two Cometto model 52M units, each having five axles and a special curved bed both to accept the round pipes but, more importantly, to keep the load as low as possible for passage through the internal tunnels where clearance at times was only 1 ft 8 ins. The dished construction meant that each axle could only be fitted with two pairs of small tyres but their carrying capacity was still 13 tonnes per line provided a speed limit of 3 mph was observed. Inching around the inside of the mountain where gradients on wet slate were as steep as 1 in 8 in places meant that there was little likelihood of the outfit being able to go faster than a walking pace, so the speed restriction was irrelevant. For the lighter pipes the trailers were used individually but for the heaviest sections which were as long as 75 ft, the trailers were joined up as a 48 ft long ten row pulled by a 19 ft 6 ins long drawbar to overcome the fore and aft load overhangs. The high manoeuvreability through all-wheel steering, plus the electro-hydraulic suspension of the trailers were crucial features of getting the linings into position. The Scammell may have done all the pushing and pulling, but the steering of the trailer was done manually through the use of a 'wander lead' plugged into the hydraulic control panel. At the point of installation the suspension was activated to raise the load 1 ft 2 ins where it was supported on jacks to leave it clear of the trailer bed as the hydraulics were lowered and the trailer withdrawn.

Whessoe were pleased with the performance of the combination as were Crane Fruehauf who were distributing the Cometto in the UK. The market was getting congested with numerous different brands of modular trailer but manufacturers realised that once a haulier opted for a particular type he was more than likely to re-order with that make to ensure continuity. There were exceptions, of course, with Econofreight using Nicholas yet inheriting Scheuerle when they merged with Magnaload-Mammoet. Pickfords also operated a mixture of modular gear but Heanor Haulage and Wrekin Roadways were two companies which had opted away from the popular Nicholas and gone for Cometto. Both were to operate examples of the 14-row girder trailer to haul concentrated loads, Wrekins' version, purchased in 1975 having a capacity of 250 tons. This company had literally stepped up a notch in the mid-'70s because being based at Telford made them ideally placed to compete for the heavy electrical traffic out of the Midlands which had so long been the domain of either Pickfords or Wynns. Such was the success of Wrekins that, in 1979, they were bought out by the Bulwark Group to join forces with Sunters and Wynns rather than fight against them. Following group policy, Wrekins transferred their preference to Nicholas running gear and as the Cometto went to Hong Kong, Britain's biggest transporter joined the Shropshire fleet. 20 rows of axles supported the 310-ton capacity girders and, with a price tag of more than £600,000, it was expected to be one of the best trailers in the world. But during its maiden journey in May 1980, whilst carrying a 258-ton GEC generator from Stafford to Manchester, cracks appeared in the trailer chassis resulting in a rather premature stoppage. These may have been rectifiable, but it did bring home that, even with the best of equipment, heavy haulage is a business which is always full of surprises.

The new 20-row Nicholas may have brought axle loadings down to around the 20 tonnes mark without the need for the air cushion equipment, but with the overall length of the outfit being 245 ft it did mean that, even with all wheel steering, there were numerous road situations which could not be negotiated. Getting in and out of customers' premises was also sometimes impossible with this type of trailer so, when conditions dictated, it was back to the old 12-row and the hoverlift, although ACE 2 had now replaced ACE 1. Gone was the blower truck which hooked up behind substituting for one of the locomotives and in its place, mounted on the back of the girder trailer, was a self-contained unit consisting of four gas turbine engines which could produce a greater cushion relief in a more efficient manner than the old system. Wrekins, Wynns and Pickfords trailers which used the ACE 2 were all fitted with suitable brackets to accept the engines, so switching the new equipment from one haulier to another was also not much of a problem.

As mentioned earlier, both Wynns and Pickfords had encountered situations where bridges had become so weakened that, even with the air cushion equipment working at full thrust, there was still a great likelihood of something like the incidents at Boroughbridge and Ardrossan occurring again.

Normally routes like this were avoided like the plague but where there was no alternative then the haulier had to improvise. The CEGB had amongst its diverse equipment a large steel construction which was used for numerous things but was mainly intended as a loading ramp for getting on and off coastal shipping. The innovative haulier's use of this ramp was to place it just above, but completely clear of, the suspect roadway completely bridging the bridge so that the outfit was able to drive across the weak bridge yet at no time touch it.

Mammoet-Econofreight needed more than the CEGB loading ramp when they delivered what was then the largest load ever moved on the roads of the UK and, had it not been for a particular wedding in July 1981 coinciding with the move, then a lot more publicity would have no doubt been attracted to it. Being able to offer a complete packaged delivery operation right from the manufacturer in Holland to its destination at the PV3 lube plant in Esso's Fawley refinery, Mammoet-Econofreight were one of the few hauliers which could be seriously considered to haul the 476-tonne, 193 ft 9 ins long, 42 ft 4 ins high distillation column. The journey commenced with the SIF vessel being lifted by crane on to a barge which took it down the canals to Rotterdam. To cross the North Sea and the English Channel, the barge was floated on to a very special ship belonging to the Mammoet fleet. Looking like a cross between a deep-sea tug and a floating dry dock, the boat was constructed so that the rear part hinged open allowing the carrying part to be flooded, thus permitting smaller vessels just to float in as though they were being swallowed up by a whale. Once the sea journey was completed, the barge was disgorged and, with no suitable dock facilities available which could handle this enormous load, the sea-going mount was simply beached on the shore line adjacent to the Fawley CEGB power station. If getting the load back on to dry land was delicate, reaching the actual surfaced roads was just plain hard work. To negotiate the 440-yard stretch of hardcore, Econofreight had constructed a 28 ft wide roadway made up of 12 in by 12 in timbers. 125 yards were laid then, as the load passed over it, the vacated timbers were moved to the front in leap-frog fashion in a move which went like clockwork. On to the roadway things were relatively straightforward although, to surmount an incline inside the power station complex, four tractors were hitched up to ensure traction on the greasy road surface. On to the public roads the two leading Scammells were uncoupled and the 5-mile haul to the refinery was completed using TRL 924H, that ageing ex-Magnaload Contractor, and underlining the Dutch presence, a 6 × 4 FTF of the Mammoet fleet. Entering the Esso refinery should have been a time when everyone was to breathe a sigh of relief but devising a method to manoeuvre the beast round the internal installations was the point where UK project manager, Toby Allin, had to work for his living. There was no way the outfit could turn the required 90-degree turn in the space available, so an AK680 mobile crane was used to lift first the front and then the rear of the vessel whilst the bogies were manoeuvred into position round the corner ensuring the remainder of the haul was just a formality.

The early '80s saw recession hit the haulage industry causing a lot of operators to tighten their belts and prune their overheads. Wrekin Roadways were to disappear as they merged with Wynns Heavy Haulage, although 1982 saw the newly-named Pickfords Industrial pass out of government ownership as the NFC became a publicly-owned company. The prospects for heavy haulage are still very exciting and closing this history with an account of a very special move, leads me to believe that whenever there are loads to be carried, there will always be men and machines able to haul the heavyweights.

This well turned out Scania belonging to Dawsons of Middlesbrough is seen on a pre-operational test run, in August 1980, running at a gross combination weight of 120 tons. This configuration of Nicholas axles combined to a gooseneck which can be adjusted to vary the imposed tractor weight must be ideal for opeators who prefer to use artics to their maximum permitted 150 tons under the Special Types General Order.

Right Peter Clemmett and *Fearnought* are seen late in January 1979 with this waste heat boiler destined for the oil terminal at Sullom Voe. The automatically steered ten row vehicle is tracking well, whilst Sunters' Titan I in articulated form is seen in convoy waiting for progress to be recommenced.

Right One of the first MAN Jumbo 150-tonners seen in this country was this 40.400 which came over as a demonstrator in the summer of 1979. Pickfords had use of it for some time as did British Nuclear Fuels, but it is seen here being used by Heanor Haulage with a Blackwood Hodge caterpillar on a King tri-axle semi-trailer en route from Hartlepool to the South Coast.

Below Econofreight of Teeside are seen hauling this 110 ft long air separation plant from Edmonton, North London, to Tilbury prior to export to Africa late in 1979. All up weight was about 110 tons, well within the 180-ton capacity of the Fiat 300 PT, a rare tractor in UK heavy haulage. The hydraulic suspension of the Nicholas bogies were put to good use in loading and unloading, an asset appreciated by lots of other operators.

Left Sunters' Titan I eases over Newport Bridge, Middlesbrough, in October 1980 with one of three silver vessels in convoy manufactured by Head Wrightsons for BP Chemicals. Bound for the docks at Hartlepool, this 170 ft long 'cigar' weighed in at 130 tonnes. Ro-ro'd up to Grangemouth, the vessels were part of the No 3 ethanol plant being constructed by Davy McKee.

Below Since the opening of Foster Wheeler Power Products in 1968, Seaton Carew has seen its fair share of abnormal load traffic, however this Sunter quartet seen on July 29 1979 must rank as the biggest convoy yet. Bill Jamieson leads with a package boiler at about 250 tons destined for Hartlepool whilst his three colleagues had a further 20 miles with their charges bound for distant Teesport.

Keeping a good lookout for low flying aircraft, Wrekins moved this 85 ft long, 125-ton pedestrian walkway in October 1980 from Terminal 1 to Terminal 2 at Gatwick airport, utilising a pair of Cometto three-axle bogies.

April 1978 saw Malcolm Johnson ease his F89 back on to dry land prior to hauling these Whesshoe lining pipes up to the new Dinorwic power station, North Wales.

Miller's mighty Mack *Bonzo Bear* takes time out from trekking up to Dinorwic to deliver this 150-ton piece of electrical equipment in the London area.

Hauling this 130-ton paper drum from South Shields to Prudhoe in early 1981, Mammoet-Econofreight covered a distance as the crow flies of no more than 15 miles. But with height and weight limits combined with ice and snow playing havoc with the route, the F89-Contractor-Scheuerle outfit took ten days to complete the haul.

Bill Jamieson rounds Churchyard roundabout in Stockton early in 1981 with this 129 tonnes vessel to be ro-ro'd out to sunnier climes. Sunters do not wish to talk too much about this load for on its first journey to the docks it fell off the trailer and had to be returned to the makers for scrutiny.

Below Although having worked hard for most of its life at the ICI salt mines in Cheshire, Econofreight still expect and get a lot from this rejuvenated Contractor. It is seen close to the end of a short haul into ICI Wilton with this impressive 104-ton vessel early in 1981.

This page Sunters Titan II, their Tractormas and two of the Contractors plus 44 rows of Nicholas running gear were needed to haul 708 tons of impressive Head Wrightson vessels. The four loads were hauled from Thornaby to Hartlepool docks in June 1982 prior to their sea journey to the USSR.

The two ex-Magnaload Volvos shepherd this accommodation module across the dock side prior to its being loaded out at Lowestoft in July 1981. At 550 tons it would be described as fairly small, it being one of four similar loads bound for the Maureen field.

Above Two of the immaculate fleet of Planthaul of Bedford having just returned from Poland with two heavy castings destined for Leeds. These were no problem to the F12-Goldhofer combinations.

Below The gas project being constructed at Mossmorran, Southern Scotland in late 1982 brought lots of work for Mammoet-Econofreight and also gave the exciting new Scammell an opportunity to show its paces. Driver Ernie Pickersgill missed his normal Volvo N10 but he couldn't fault the S24 for sheer strength, yet even he was taken aback when the manufacturer's engineers asked him to stop the outfit on a 1 in 10 incline at an all up weight of 286 tons and then restart it. The big 350 Cummins and Brockhouse torque converter driving through Spicer gearbox and Michelin sand type tyres didn't even sigh and moved off as if they were on the flat.

Left The N10 is not a common machine in UK heavy haulage, although Econofreight find that the fitting of a torque converter more than compensates for the lack of power in the 290 engine. The Volvo-Nicholas combination is seen on a short haul across Cleveland in March 1982, all up weight !16 tons, overall length 143 ft.

Below Mammoet-Econofreight hauling in July 1981 what was then the heaviest load moved on roads in the UK. At 476 tonnes, 193 ft 9 ins long and 42 ft 4 ins high, the vessel needed four tractors to ensure traction up this incline on the private roads inside the Fawley power station. The 5-mile road haul was completed using the ex-Magnaload Contractor and an FTF of the Mammoet fleet.

Right Although this enormous monster was termed a mobile crane, when it came to moving from one end of the Port of London to another it was far more efficient to use a heavy haulier. Mammoet-Econo-freight used ten rows of Nicholas running gear plus two of their strongest Volvos as a pre-requisite of the all up weight being in excess of 350 tons.

Below Coming to the Sunter fleet in late 1981 was this French-built Tractormas. One of its first road hauls was the carriage of this 120-ton Head Wrightson kiln section from Thornaby to Avonmouth in May 1982.

CJB
CJB
WHESSOE
WHESSOE
WHESSOE